MALESIAN SEED PLANTS

Volume 1 — Spot-characters

An aid for identification of families and genera

M. M. J. van Balgooy

Rijksherbarium / Hortus Botanicus
Leiden — 1997

ISBN 90-71236-31-5

Cover design: Richard T. van Balgooy, The Hague, The Netherlands
Lay-out::   Emma E. van Nieuwkoop, Oegstgeest, The Netherlands
Print: Grafische Vormgeving Kanters, Sliedrecht, The Netherlands
Distribution: Backhuys Publishers, P. O. Box 321, 2300 AH Leiden, The Netherlands

CONTENTS

PREFACE

The Malesian region, which includes the nation states of Brunei Darussalam, Indonesia, Malaysia, Papua New Guinea, the Philippines, and Singapore, harbours one of the richest and most diverse tropical floras of the world. Currently the flowering plants are estimated to include about 36,000 species, of which only 15% have been recently revised and treated in *Flora Malesiana*. Of the other 85% information is scattered in specialized literature or completely wanting. Meanwhile there is an increasingly urgent need for biodiversity expertise that will enable identification of the remaining botanical resources of Malesia, both the endangered primary vegetations as well as the rapidly expanding, yet disturbed, species-rich secondary ecosystems.

The recognition to which family a plant belongs, is the first, and often most important step to species identification, and consequently to all scientific and practical information on an unknown plant. Identification keys to the whole Malesian flora are not available. Regional works like the *Tree Flora of Malaya* and the *Flora of Java* fill important gaps in this respect, but there is a widely felt need for user-friendly identification means for the whole of Malesia.

The series '*Malesian Seed Plants*' by Dr. M.M.J. van Balgooy for the first time offers such a tool. Having spent his youth among the plants of Java, Dr. van Balgooy developed his unique botanical knowledge under the guidance of Prof. Dr. C.G.G.J. van Steenis, the founder of the Flora Malesiana project. In this series the author shares part of that knowledge in original accounts of all Malesian plant families, and in lists of spot-characters that easily lead to family or genus recognition. The descriptions are designed as portraits, highlighting the occurrence of these spot-characters: they are not intended as comprehensive and comparable descriptions in the classical sense. That remains the realm of *Flora Malesiana* and other floras. Numerous illustrations are included to facilitate the use of the 'spot-characters' and 'portraits', and technical terms have been avoided where possible, or are clearly explained.

On behalf of the Foundation Flora Malesiana I express the hope that many professionals and amateur botanists will use these books to familiarize themselves with the rich Malesian flora. Ultimately, it is hoped that this book will contribute to the protection and rational utilization of the botanical resources of the biodiversity megacentre of the Malesian region.

Bogor, January 1997

Prof. Dr. Mien A. Rifai
Chairman of the Board of the Foundation Flora Malesiana
Assistant Minister of State for Research and Technology
Republic of Indonesia

INTRODUCTION

Historical background

The concept of this book has taken some decades to acquire a definite shape. About thirty years ago I started attending the so-called pre-identification sessions of the then director of the Rijksherbarium, Prof. C.G.G.J. van Steenis († 14 May 1986) and Dr. R.C. Bakhuizen van den Brink Jr. († 1 May 1987), the co-author of the 'Flora of Java' (Backer & Bakhuizen van den Brink, 1963–1968). They spent one day a week going through all incoming material, checking the identifications on the labels and identifying unnamed specimens. The plants not recognized at first sight were put aside for further scrutiny. It was usually Rein Bakhuizen who enjoyed cracking the hard nuts. In the case of identifying trees Van Steenis and Bakhuizen very often relied on Mr. F.H. Hildebrand († 7 July 1975), who, as a former forester of the Forest Research Institute in Bogor, had a vast knowledge of Malesian tree species.

These sessions were quite unforgettable, although I must admit that the first few years were pretty rough. The two went through the piles of material like a whirlwind, each trying to beat the other in naming the plants. The only break I had was when they had an argument, or when I put in a silly question, such as: "How do you know?"

In the beginning I was quite overawed by the seemingly unlimited knowledge of the three gentlemen and I was absolutely convinced that it was impossible for me to store away so many plant names and characters into my memory. Each of the three had his own method of memorizing plants. Van Steenis was in the habit of jotting down on small scraps of paper all striking characters he came across, to enter them later in a kind of record book. Hildebrand used to prepare sketches of all plants he identified. These pencil drawings have been assembled in nineteen volumes which are kept in the library of the Rijksherbarium, where I still consult them regularly. Bakhuizen used to go through lists of genera of the various families. Whenever he encountered a name which he could not associate with a clear mental picture of the plant he went to the collection to see what it looked like. I myself used to write down everything I heard during conversations with all three seniors.

Several of my young colleagues who regularly attended the pre-identification sessions also complained that it was very difficult to remember the numerous family and genus characters. To aid our memory Van Steenis and Bakhuizen compiled the constant characters of some 100 Malesian flowering plant families (in Dutch). Some twenty years ago Van Steenis entrusted me with his record book of spot-characters. Since then I have finished family characterizations for all but a few herbaceous families and have more than doubled the number of spot-characters. A booklet by Dr. P.J.M. Maas, "Neotropische flora van A tot Z", describing all Neotropical families, has further helped to give the present effort its definite shape. This work has also appeared in an English translation (Maas & Westra, 1993).

Until today I am still adding to the lists of spot-characters and one might wonder if the present publication is not premature. However, several colleagues have pleaded with me to make available the knowledge built up over many years, so that it can serve as a tool in plant naming at the various institutes in the area. The data may eventually also be used to generate a computer key.

After consultation with various colleagues it was decided to bring all this information together under the general title 'Malesian Seed Plants' in three books: Volume 1 'Spot-characters', Volume 2 'Portraits of tree families', and Volume 3 'Portraits of non-tree families'. Volumes 2 and 3 will contain brief characterizations of the various seed plant families, 'portraits'. Each volume will be published separately.

About this book

Volume 1 of the series 'Malesian Seed Plants' contains lists of spot-characters most of which, with some training, can be easily observed in herbarium material. These characters have been arranged in a more or less logical way, e.g., characters of the stem, the leaves, the flowers, the fruits, etc. Each spot-character is explained and, where appropriate or possible, illustrated. As stated above, the lists are updated until the last moment before publication, but some are still desperately incomplete. Moreover, many spot-characters I use when identifying plants are difficult to put in words and have not been listed. These include shades, colour and texture of dried material, 'feel' and smell. Also not listed are many field characters such as those of slash and bark, crown-shape and architecture, because they are of little use to identify herbarium material.

The lists of spot-characters also contain a few non-Malesian taxa which I happened to have come across, but no attempt has been made at completeness for these extra-Malesian taxa. They are not mentioned under the heading 'spot-characters' in the family portraits of Volumes 2 and 3. Every entry has been checked in the herbarium and I have not relied on data from literature.

Although the text of this volume was finished in 1994 (and updated until end 1996), publication has unfortunately been long delayed due to problems with my health. In the meantime an interesting identification manual has been published (Keller, 1996), but I have not been able to test it, neither to incorporate data from this book into mine.

By publishing this book for the benefit of the botanical community the deficiencies can be revealed and hopefully corrected in a future revised edition. Users of this volume are kindly invited to send corrections and additions to the Rijksherbarium / Hortus Botanicus, P.O. Box 9514, 2300 RA Leiden, The Netherlands.

Literature

Keller, R. 1996. Identification of tropical woody plants in the absence of flowers and fruits. A field guide. Basel, etc.
Maas, P.J.M. & L.Y.Th. Westra. 1993. Neotropical plant families. A concise guide to families of vascular plants in the neotropics. Koenigstein / Champaign.

Abbreviations and signs

(AS) behind a name indicates that the taxon is only known from Asia
(Au) taxon only known from Australia
(P) taxon only known from the Pacific
p.p. the spot-character is not always visible or is found only in part of the taxon
* the taxon is represented in Malesia by introduced species only

ACKNOWLEDGEMENTS AND DEDICATION

Several people have contributed to the completion of this book. Attempting to name them all holds the risk of forgetting some. Therefore, let it suffice to mention by name just a few who have substantially helped to improve the text. Mr. K.M. Kochummen has supplied me with additional spot-characters. Dr. P.F. Stevens and Mr. M.J.E. Coode critically read the text and suggested many corrections. A great many colleagues both from abroad and from the Rijksherbarium have given advice, information and encouragement. Prof. C. Kalkman critically read the final version of the manuscript. I am grateful to Prof. P. Baas who convinced me to overcome reluctance against publishing this work.

My assistant, Mr. L.B.T. Kostermans, has helped to type and retype the various versions of this book and Ms. E.E. van Nieuwkoop gave the finishing touches in the lay-out with all the skills we have come to expect from her. Ms. J.R. Kruijer helped to select the illustrations and Mr. J.H. van Os prepared many of them for publication. I am indebted to various persons and institutions for the permission to reproduce drawings; their collaboration is acknowledged with the illustrations.

I am particularly obliged to my former teachers in the art of identification, Prof. Van Steenis and Dr. Bakhuizen van den Brink, who have taught me almost all they knew about Malesian plant taxonomy. It is therefore to their memory that I dedicate this book in gratitude.

Leiden, 1997   M.M.J. van Balgooy

LIST OF SPOT-CHARACTERS

8

HABIT (characters 1–13)

1. Cushion plants — Fig. 1

These are plants that form compact masses, often in the form of a cushion. This habit is very common in the South American Andes and in New Zealand; in Malesia this habit is almost confined to alpine vegetation on the highest mountains, especially in New Guinea; examples *Centrolepis* and *Rhododendron saxifragoides*.

Taxon	Family	Taxon	Family
Astelia p.p.	Liliac.	*Oreobolus*	Cyp.
Centrolepis	Centr.	*Oreomyrrhis* p.p.	Umb.
Cerastium p.p.	Caryoph.	*Plantago* p.p.	Plant.
Coprosma archboldiana	Rub.	*Pleiocraterium gentianifolia*	Rub.
Danthonia p.p. (*Monostachya*)	Gram.	*Potentilla* p.p.	Rosac.
Drosera p.p.	Dros.	*Rhamphogyne*	Comp.
Eriocaulon p.p.	Erioc.	*Rhododendron caespitosum*	Eric.
Gaimardia	Centr.	*Rhododendron saxifragoides*	Eric.
Gentiana p.p.	Gent.	*Sagina* p.p.	Caryoph.
Geranium p.p.	Geran.	*Trachymene* p.p.	Umb.
Isachne p.p.	Gram.	*Trigonotis* p.p.	Borag.
Lactuca p.p.	Comp.	*Xyris* p.p.	Xyr.
Lepidium p.p. (*Papuzilla*)	Cruc.		

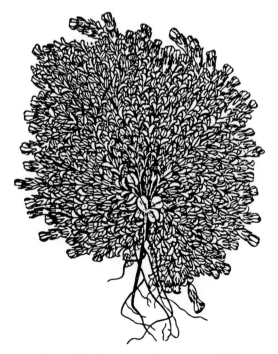

Figure 1. Cushion plants – *Gentiana quadrifaria.*

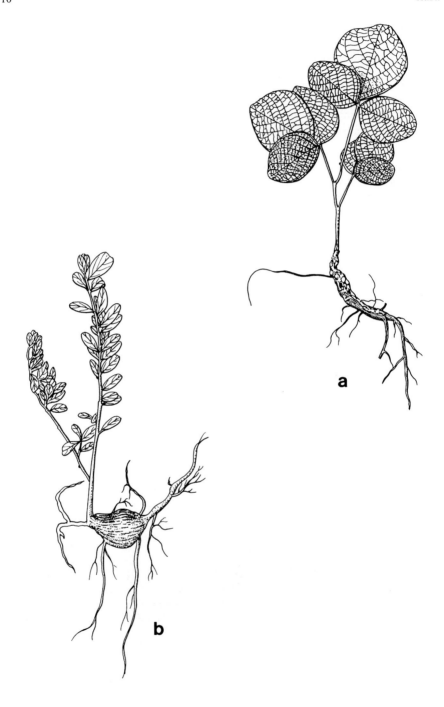

Figure 2. Swollen stems – a. *Butea monosperma*; b. *Vaccinium lucidum*; c. *Anthorrhiza chrysacantha* (see next page).

2. Swollen stems — Fig. 2

Plants with gouty or swollen stems; in some species, such as the *Hydnophytinae,* they are hollow and inhabited by ants, in others they store water, as in some *Impatiens.* In other taxa they consist of underground parts (lignotubers) by which the plant survives fires or severe droughts, a rare phenomenon in Malesia but common in Australia.

Taxon	*Family*	*Taxon*	*Family*
Agapetes p.p.	Eric.	*Myrmecodia*	Rub.
Anthorrhiza	Rub.	*Myrmephytum*	Rub.
Brachychiton p.p. (Au)	Sterc.	*Neoalsomitra* p.p.	Cuc.
Butea monosperma p.p.	Leg.	*Neptunia oleracea*	Leg.
Cissus p.p.	Vit.	*Pachycentria*	Melast.
Hydnophytum	Rub.	*Pachynema* (Au)	Dill.
Impatiens p.p.	Bals.	*Planchonia* p.p. (Au)	Lecyth.
Jatropha *	Euph.	*Pogonanthera*	Melast.
Leguminosae p.p.	Leg.	*Premna* p.p. (*Pygmaeopremna*)	Verb.
Leucas p.p.	Lab.	*Vaccinium* p.p.	Eric.
Medinilla p.p.	Melast.	*Vitex* p.p.	Verb.

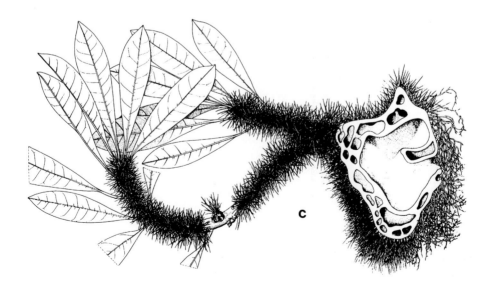

Figure 3. Monocarpic plants – *Corypha elata.*

3. Monocarpic plants — Fig. 3

These are perennial plants that produce one inflorescence after which they die, as for instance in *Metroxylon*.

Taxon	Family	Taxon	Family
Agave *	Liliac.	*Harmsiopanax*	Aral.
Bambusoideae p.p.	Gram.	*Metroxylon*	Palm.
Corypha	Palm.	*Strobilanthes* p.p.	Acanth.

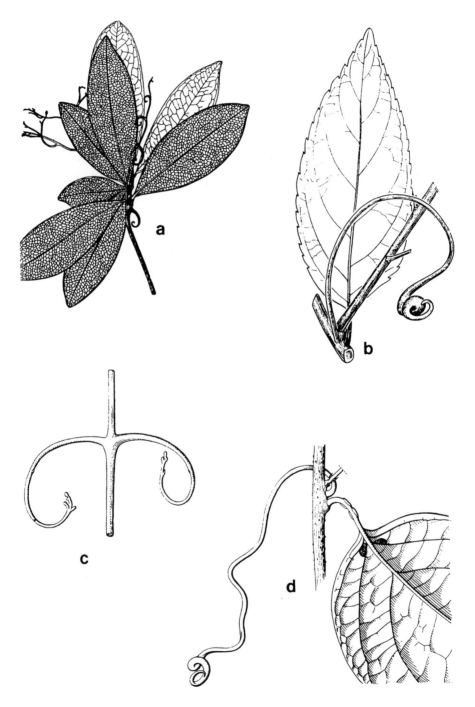

Figure 4. Climbers with hooks / tendrils – a. *Ancistrocladus tectorius*; b. *Lophopyxis maingayi*; c. *Enkleia malaccensis*; d. *Hollrungia aurantioides*.

4. Climbers with hooks/tendrils — Fig. 4, 6b

Plants that climb by means of some special aid. This can be in the form of prehensile tendrils, such as in *Cucurbitaceae* or *Vitaceae*; special branches that grow around footholds, such as seen in many climbing *Annonaceae*; yet others have curved woody hooks such as found in *Uncaria* and many rattans.

Taxon	Family	Taxon	Family
Acacia p.p.	Leg.	*Lathyrus* *	Leg.
Adenia	Passifl.	*Lophopyxis*	Loph.
Ampelocissus	Vit.	*Luvunga*	Rut.
Ampelopsis	Vit.	*Lysiphyllum*	Leg.
Ancistrocladus	Ancistr.	*Maclura*	Morac.
Antigonon *	Polygon.	*Melodorum*	Annon.
Artabotrys	Annon.	*Mimosa* *	Leg.
Bauhinia p.p.	Leg.	*Myrialepis*	Palm.
Bignoniaceae p.p.*	Bign.	*Naravelia*	Ranunc.
Bougainvillea *	Nyctag.	*Nepenthes*	Nepenth.
Bracteolanthus	Leg.	*Nothocissus*	Vit.
Caesalpinia p.p.	Leg.	*Olax*	Olacac.
Calamus	Palm.	*Omphalea* p.p.	Euph.
Callerya p.p. (*Whitfordiodendron*)	Leg.	*Partenocissus*	Vit.
Calospatha	Palm.	*Passiflora*	Passifl.
Canthium p.p.	Rub.	*Petraeovitex*	Verb.
Capparis p.p.	Capp.	*Philbornea*	Linac.
Cardiospermum *	Sapind.	*Pisonia (aculeata)*	Nyctag.
Cayratia	Vit.	*Pisum* *	Leg.
Ceratolobus	Palm.	*Plectocomia*	Palm.
Cissus	Vit.	*Plectocomiopsis*	Palm.
Clerodendrum p.p.	Verb.	*Pogonotium*	Palm.
Cucurbitaceae p.p.	Cuc.	*Polyporandra* p.p.	Icacin.
Daemonorops	Palm.	*Pterisanthes*	Vit.
Dalbergia p.p.	Leg.	*Quisqualis*	Combr.
Enkleia	Thym.	*Randia* s.l. p.p.	Rub.
Entada p.p.	Leg.	*Rauwenhoffia*	Annon.
Erythropalum	Olacac.	*Retispatha*	Palm.
Flagellaria	Flag.	*Rubus*	Rosac.
Friesodielsia	Annon.	*Sageretia* p.p.	Rhamn.
Gloriosa	Liliac.	*Smilax*	Liliac.
Gouania	Rhamn.	*Smythea* p.p.	Rhamn.
Harrisonia	Simar.	*Solanum* p.p.	Solan.
Heterosmilax	Liliac.	*Strychnos* p.p.	Logan.
Hollrungia	Passifl.	*Tetrastigma*	Vit.
Hugonia	Linac.	*Toddalia*	Rut.
Illigera	Hern.	*Uncaria*	Rub.
Indorouchera	Linac.	*Uvaria*	Annon.
Iodes	Icacin.	*Ventilago*	Rhamn.
Jasminum p.p.	Oleac.	*Vitis*	Vit.
Korthalsia	Palm.	*Willughbeia*	Apoc.
Lantana p.p.*	Verb.	*Zanthoxylum* p.p.	Rut.
Lasiobema	Leg.	*Zizyphus*	Rhamn.

Figure 5. Climbers without hooks / tendrils – a. *Dioscorea bulbifera* (twining left); b. *Dioscorea laurifolia* (twining right); c. *Rhaphidophora korthalsii* (Arac.).

5. Climbers without hooks / tendrils — Fig. 5, 6a

Plants climbing by means of a twining stem, such as many *Leguminosae* and *Menispermaceae*, or with adhesive roots as many *Araceae*, but without specialised climbing organs.

Taxon	Family	Taxon	Family
Acacia p.p.	Leg.	*Cyrtandra* p.p.	Gesn.
Actinidia	Actin.	*Dalbergia* p.p.	Leg.
Aeschynanthus	Gesn.	*Dalechampia*	Euph.
Agalmyla	Gesn.	*Deeringia*	Amaran.
Aganope	Leg.	*Derris*	Leg.
Aganosma	Apoc.	*Desmos* p.p.	Annon.
Agatea	Viol.	*Dichapetalum* p.p.	Dichap.
Agelaea	Connar.	*Dinochloa*	Gram.
Aidiopsis	Rub.	*Dioclea*	Leg.
Airyantha	Leg.	*Dioscorea*	Diosc.
Alyxia	Apoc.	*Diplectria*	Melast.
Anomanthodia	Rub.	*Dipodium scandens*	Orch.
Anomianthus	Annon.	*Ecdysanthera*	Apoc.
Anredera	Basell.	*Ellipeia*	Annon.
Araceae p.p.	Arac.	*Embelia* p.p.	Myrsin.
Aristolochia	Arist.	*Entada* p.p.	Leg.
Asclepiadaceae p.p.	Asclep.	*Epigynum*	Apoc.
Aspidopteris	Malp.	*Erycibe* p.p.	Conv.
Bauhinia p.p.	Leg.	*Euonymus* p.p.	Celastr.
Berchemia p.p.	Rhamn.	*Eustrephus*	Liliac.
Bowringia	Leg.	*Faradaya*	Verb.
Bridelia p.p.	Euph.	*Ficus* p.p.	Morac.
Byttneria	Sterc.	*Fissistigma*	Annon.
Caesalpinia p.p.	Leg.	*Freycinetia*	Pand.
Callerya (Whitfordiodendron)	Leg.	*Galeola*	Orch.
Cansjera	Opil.	*Garcinia* (SAN 77272)	Gutt.
Cardiopteris	Card.	*Gardneria*	Logan.
Cassytha	Laur.	*Geitonoplesium*	Liliac.
Celastrus	Celastr.	*Glossocarya*	Verb.
Clematis	Ranunc.	*Gnetum* p.p.	Gnet.
Clitorea	Leg.	*Gynochthodes*	Rub.
Cnesmone	Euph.	*Gynopachis*	Rub.
Cnestis	Connar.	*Hibbertia* p.p.	Dill.
Coelospermum	Rub.	*Hieris*	Bign.
Combretum	Combr.	*Hiptage*	Malp.
Congea	Verb.	*Hosea*	Verb.
Connarus	Connar.	*Hymenopyramis* (As)	Verb.
Coptosapelta	Rub.	*Illigera*	Hern.
Crawfurdia	Gent.	*Ipomoea* p.p.	Conv.
Creochiton	Melast.	*Ischnocarpus*	Apoc.
Croton caudatus p.p.	Euph.	*Jacquemontia*	Conv.
Cuscuta	Conv.	*Jasminum* p.p.	Oleac.
Cyathostemma	Annon.		

(5. Climbers without hooks / tendrils, continued)

Taxon	Family	Taxon	Family
Kadsura	Schis.	*Phytocrene*	Icacin.
Kunstleria	Leg.	*Plagiopteron* (As)	Plag.
Leuconotis	Apoc.	*Polygala* p.p.	Polygal.
Linostoma	Thym.	*Polyporandra* p.p.	Icacin.
Loeseneriella	Celastr.	*Porana*	Conv.
Lonicera	Caprif.	*Pottsia*	Apoc.
Lucinaea	Rub.	*Premna* p.p.	Verb.
Macrolenes	Melast.	*Psychotria* p.p.	Rub.
Macropsychanthus	Leg.	*Pterococcus*	Euph.
Maesa p.p.	Myrsin.	*Pterolobium* *	Leg.
Malaisia	Morac.	*Pueraria*	Leg.
Mastersia	Leg.	*Pycnospora*	Leg.
Maurandya *	Scroph.	*Pyramidanthe*	Annon.
Medinilla p.p.	Melast.	*Pyrenacantha*	Icacin.
Megistostigma	Euph.	*Quisqualis*	Combr.
Melodinus	Apoc.	*Racemobambos*	Gram.
Menispermaceae p.p.	Menisp.	*Rhipogonum*	Liliac.
Merremia	Conv.	*Rhynchosia*	Leg.
Micrechites	Apoc.	*Rhyssopterys*	Malp.
Mikania	Comp.	*Rourea*	Connar.
Millettia p.p.	Leg.	*Roureopsis*	Connar.
Miquelia	Icacin.	*Sabia*	Sab.
Mitrella	Annon.	*Salacia*	Celastr.
Monarthrocarpus	Leg.	*Sarcodum*	Leg.
Morinda p.p.	Rub.	*Sarcostigma*	Icacin.
Mucuna	Leg.	*Scaevola oppositifolia*	Good.
Muehlenbeckia	Polygon.	*Schisandra*	Schis.
Mussaenda p.p.	Rub.	*Securidaca*	Polygal.
Myxopyrum	Oleac.	*Smythea* p.p.	Rhamn.
Nastus	Gram.	*Spatholirion*	Comm.
Neodissochaeta	Melast.	*Spatholobus*	Leg.
Neosepicaea	Bign.	*Sphenodesme*	Verb.
Nyctocalos	Bign.	*Stemona*	Stem.
Omphalea p.p.	Euph.	*Strongylodon*	Leg.
Operculina	Conv.	*Strophanthus*	Apoc.
Opilia	Opil.	*Symphorema*	Verb.
Pachystylidium	Euph.	*Tecomanthe*	Bign.
Padbruggea	Leg.	*Tephrosia*	Leg.
Paederia	Rub.	*Tetracera*	Dill.
Palmeria	Monim.	*Thunbergia*	Acanth.
Pandorea	Bign.	*Tournefortia*	Borag.
Parsonsia	Apoc.	*Trimenia macrura*	Trim.
Parvatia (As)	Lard.	*Tristellateia*	Malp.
Pegia	Anac.	*Urceola*	Apoc.
Petraeovitex	Verb.	*Vanilla*	Orch.
Phylacium	Leg.	*Vernonia* p.p.	Comp.
Phyllanthus reticulatus	Euph.	*Vigna*	Leg.

6. Climbers with opposite leaves — Fig. 6

Plants climbing, with or without tendrils or hooks, with opposite leaves, e.g. many *Apocynaceae*, *Asclepiadaceae* and *Bignoniaceae*.

Taxon	Family	Taxon	Family
Aeschynanthus p.p.	Gesn.	*Hiptage*	Malp.
Agalmyla	Gesn.	*Hosea*	Verb.
Aganosma	Apoc.	*Hydrangea* p.p.	Sax.
Aidiopsis	Rub.	*Hymenopyramis* (As)	Verb.
Allaeophania	Rub.	*Iodes*	Icacin.
Alyxia p.p.	Apoc.	*Ischnocarpus*	Apoc.
Anomanthodia	Rub.	*Jasminum* p.p.	Oleac.
Aphaenandra	Rub.	*Leuconotis*	Apoc.
Artia	Apoc.	*Leviera* p.p.	Monim.
Asclepiadaceae p.p.	Asclep.	*Linostoma*	Thym.
Aspidopteris	Malp.	*Lonicera*	Caprif.
Caesalpinia oppositifolia	Leg.	*Lucinaea*	Rub.
Calycopteris	Combr.	*Macrolenes*	Melast.
Canthium p.p.	Rub.	*Maurandya* *	Scroph.
Catanthera	Melast.	*Medinilla* p.p.	Melast.
Chilocarpus	Apoc.	*Melodinus*	Apoc.
Chonemorpha	Apoc.	*Micrechites*	Apoc.
Clematis	Ranunc.	*Mikania*	Comp.
Coelospermum	Rub.	*Morinda* p.p.	Rub.
Combretum	Combr.	*Mussaenda* p.p.	Rub.
Congea	Verb.	*Myxopyrum*	Oleac.
Coptosapelta	Rub.	*Naravelia*	Ranunc.
Crawfurdia	Gent.	*Neodissochaeta*	Melast.
Creochiton	Melast.	*Neosepicaea*	Bign.
Cyrtandra p.p.	Gesn.	*Paederia*	Rub.
Dioscorea p.p.	Diosc.	*Palmeria*	Monim.
Diplectria	Melast.	*Pandorea*	Bign.
Dissochaeta	Melast.	*Parabarium*	Apoc.
Ecdysanthera	Apoc.	*Parameria*	Apoc.
Enkleia	Thym.	*Parsonsia*	Apoc.
Epigynum	Apoc.	*Petraeovitex*	Verb.
Euonymus p.p.	Celastr.	*Plagiopteron* (As)	Plag.
Fagraea p.p.	Logan.	*Polyporandra*	Icacin.
Faradaya	Verb.	*Pottsia*	Apoc.
Ficus p.p.	Morac.	*Premna* p.p.	Verb.
Garcinia (SAN 77272)	Gutt.	*Psychotria* p.p.	Rub.
Gardneria	Logan.	*Quisqualis*	Combr.
Gelsemium p.p.	Logan.	*Rhynchodia*	Apoc.
Glossocarya	Verb.	*Rhyssopterys*	Malp.
Gnetum p.p.	Gnet.	*Salacia* p.p.	Celastr.
Gynochthodes	Rub.	*Saritaea* *	Bign.
Gynopachis	Rub.	*Scaevola oppositifolia*	Good.
Hieris	Bign.	*Sphenodesme*	Verb.

(6. Climbers with opposite leaves, continued)

Taxon	Family	Taxon	Family
Strophanthus	Apoc.	*Trichopus*	Diosc.
Strychnos p.p.	Logan.	*Trimenia macrura*	Trim.
Symphorema	Verb.	*Tristellateia*	Malp.
Tecomanthe	Bign.	*Uncaria*	Rub.
Thunbergia	Acanth.	*Urceola*	Apoc.
Tournefortia p.p.	Borag.	*Urnularia*	Apoc.
Trachelospermum	Apoc.	*Willughbeia*	Apoc.

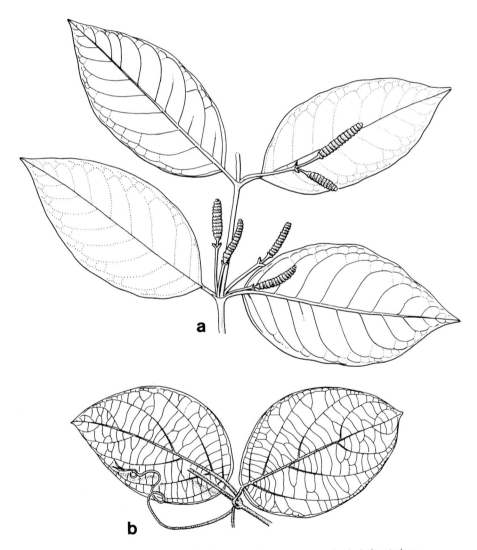

Figure 6. Climbers with opposite leaves – a. *Gnetum gnemonoides*; b. *Iodes cirrhosa*.

7. Echlorophyllose plants — Fig. 7 (see also Fig. 8a and 11a, p. 22 and 28)

Plants devoid of chlorophyll: either saprophytes such as *Triuridaceae* and the orchid genera *Aphyllorchis* and *Lecanorchis*, or holoparasites such as *Rafflesia* and *Balanophora*.

Taxon	*Family*
Aeginetia	Orob.
Andresia	Eric.
Aphyllorchis	Orch.
Balanophora	Balanoph.
Burmannia p.p.	Burm.
Cassytha	Laur.
Christisonia	Orob.
Corsia	Cors.
Corybas p.p.	Orch.
Cotylanthera	Gent.
Cuscuta	Conv.
Cystorchis	Orch.
Didymoplexiella (As)	Orch.
Didymoplexis	Orch.
Epipogum	Orch.
Epirixanthes	Polygal.
Eulophia	Orch.
Exorhopala	Balanoph.
Galeola	Orch.
Gastrodia	Orch.
Gymnosiphon	Burm.
Hypopithys (As)	Eric.
Langsdorffia	Raffl.
Lecanorchis	Orch.
Mitrastemma	Raffl.
Monotropastrum	Eric.
Pachystoma	Orch.
Petrosavia	Liliac.
Rafflesia	Raffl.
Rhizanthes	Balanoph.
Rhopalocnemis	Balanoph.
Sapria	Orch.
Sciaphila	Triur.
Thismia	Burm.

Figure 7. Echlorophyllose plants – *Sciaphila densiflora*.

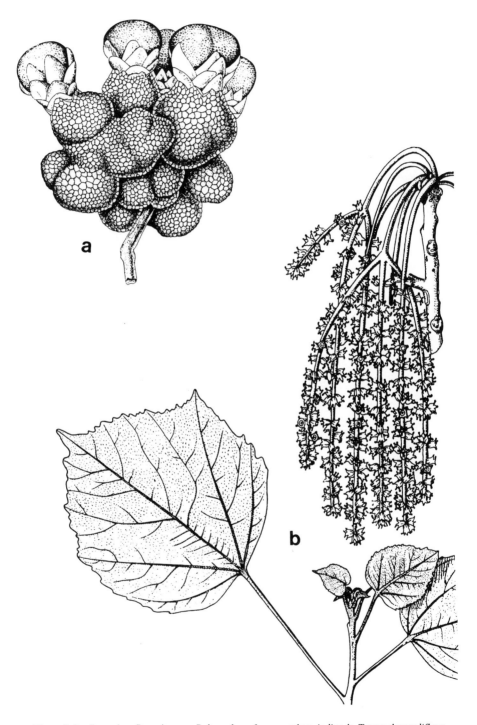

Figure 8. Leafless when flowering – a. *Balanophora fungosa* subsp. *indica*, b. *Tetrameles nudiflora*.

8. Leafless when flowering — Fig. 8

Plants without leaves when in flower, either because they are always leafless such as *Taeniophyllum* or *Sarcostemma* (l), or plants dropping their leaves before flowering such as *Tetrameles* and *Firmiana*.

Taxon	Family	Taxon	Family
Amorphophallus	Arac.	*Lannea* *	Anac.
Balanophoraceae	Balan.	*Mayodendron* p.p. (As)	Bign.
Bombax	Bomb.	*Parishia* p.p.	Anac.
Catunaregam	Rub.	*Parkia* p.p.	Leg.
Chiloschista (l)	Orch.	*Placellaria* (l) p.p.	Sant.
Combretum	Combr.	*Premna* p.p.	Verb.
Crepis p.p.	Comp.	*Radermachera* p.p.	Bign.
Dalbergia p.p.	Leg.	*Rafflesiaceae*	Raffl.
Dillenia p.p.	Dill.	*Remusatia*	Arac.
Dipterocarpus p.p.	Dipt.	*Sarcostemma* (l)	Asclep.
Erythrina p.p.	Leg.	*Shorea* p.p.	Dipt.
Firmiana	Sterc.	*Sterculia* p.p.	Sterc.
Flacourtia p.p.	Flac.	*Stereospermum*	Bign.
Gardenia p.p	Rub.	*Symphorema* p.p.	Verb.
Garuga	Burs.	*Taeniophylum* (l)	Orch.
Gmelina p.p.	Verb.	*Tectona*	Verb.
Hildegardia	Sterc.	*Terminalia* p.p.	Combr.
Hymenodictyon	Rub.	*Tetrameles*	Datisc.
Itoa	Flac.		

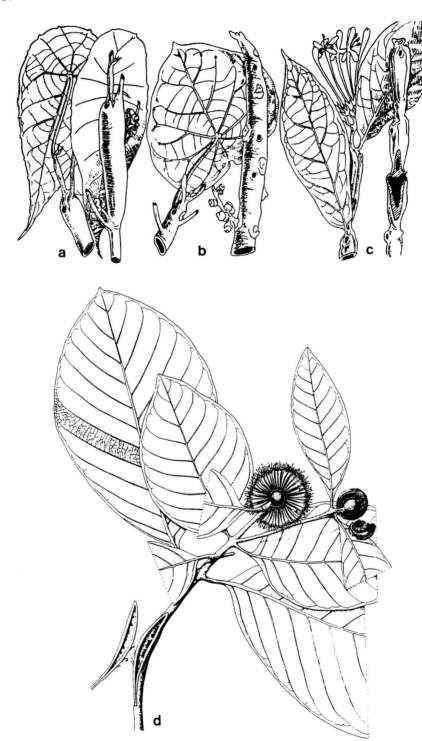

9. Ant plants — Fig. 9 (see also Fig. 2a, p. 10)

Plants that have special constructions providing housing for ants. Best known are the members of the *Hydnophytinae* with stems provided with a labyrinth of holes. Other well-known examples of ant-inhabited plants are species of *Endospermum* and *Macaranga*.

Taxon	Family	Taxon	Family
Acacia p.p.	Leg.	*Hydnophytum*	Rub.
Alpinia domatifera	Zing.	*Kibara* p.p.	Monim.
Amylotheca formicaria	Loranth.	*Korthalsia* p.p.	Palm.
Anthorrhiza	Rub.	*Macaranga* p.p.	Euph.
Archidendron aruense	Leg.	*Myristica myrmecophila*	Myrist.
Chisocheton myrmecophilus	Meliac.	*Myristica subalulata*	Myrist.
Clerodendrum p.p.	Verb.	*Myrmecodia*	Rub.
Dischidia p.p.	Asclep.	*Myrmeconauclea*	Rub.
Drypetes myrmecophila	Euph.	*Neonauclea* p.p.	Rub.
Drypetes pendula	Euph.	*Nepenthes bicalcarata*	Nepenth.
Elaeocarpus myrmecophilus	Elaeoc.	*Piper microphyllum*	Piper.
Endospermum p.p.	Euph.	*Psychotria myrmecophila*	Rub.
Euroschinus	Anac.	*Rinorea javanica*	Viol.
Ficus p.p.	Morac.	*Saurauia myrmecoidea*	Actin.
Harpullia myrmecophila	Sapind.	*Semecarpus aruensis*	Anac.
Homalanthus fastuosus	Euph.	*Steganthera* p.p.	Monim.
Hoya p.p.	Asclep.	*Zanthoxylum myriacanthum*	Rut.

←

Figure 9. Ant plants – a. *Macaranga caladiifolii*; b. *Endospermum moluccanum*; c. *Clerodendrum fistulosum*; d. *Neonauclea superba*.

Figure 10. Schopfbäume – *Cycas rumphii* (drawn by Mrs. R.S. Keng).

10. Schopfbäume — Fig. 10

A German term for trees more or less shaped like an umbrella: unbranched or little branched and usually with large, crowded leaves. *Cycas* is a good example.

Taxon	Family	Taxon	Family
Agrostistachys	Euph.	*Meliaceae* p.p.	Meliac.
Anakasia	Aral.	*Osmoxylon* p.p.	Aral.
Barringtonia p.p.	Lecyth.	*Palmae* p.p.	Palm.
Carica *	Caric.	*Pandanus*	Pand.
Cordyline	Liliac.	*Rubiaceae* p.p.	Rub.
Cycas	Cycad.	*Sararanga*	Pand.
Dracaena	Liliac.	*Schuurmansia*	Ochn.
Eurycoma	Simar.	*Semecarpus* p.p.	Anac.
Harmsiopanax	Aral.	*Sterculia* p.p.	Sterc.
Jagera	Sapind.	*Tapeinosperma*	Myrsin.
Leea p.p.	Leeac.		

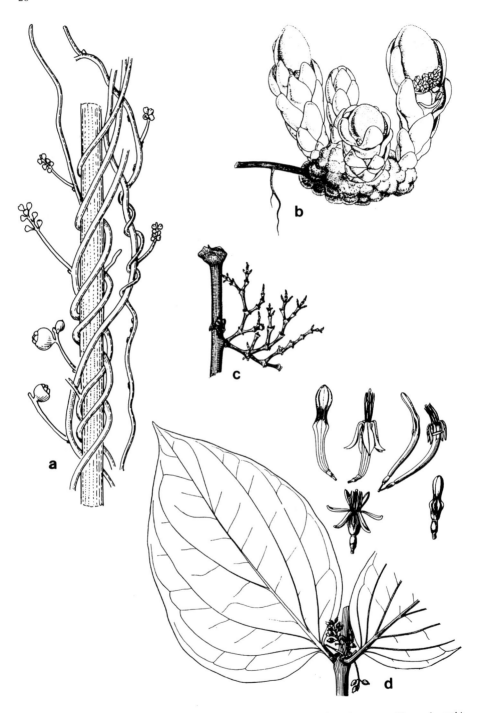

Figure 11. Parasites – a. *Cassytha filiformis*; b. *Balanophora fungosa* subsp. *fungosa*; c. *Viscum loranthi*; d. *Macrosolen curvinervis*.

11. Parasites — Fig. 11 (see also Fig. 8a, p. 22)

Plants depending in part or completely on other plants for their nutrients. Examples of the first are *Loranthaceae* and *Santalaceae,* of the second *Balanophoraceae* and *Rafflesiaceae*, i.e. holoparasites (h).

Taxon	Family	Taxon	Family
Aeginetia (h)	Orob.	*Lepeostegeres*	Loranth.
Amyema	Loranth.	*Lepidaria*	Loranth.
Amylotheca	Loranth.	*Lepidella*	Loranth.
Balanophora (h)	Balanoph.	*Loxanthera*	Loranth.
Barathranthus	Loranth.	*Macrosolen*	Loranth.
Buchnera	Scroph.	*Mitrastemma* (*Mitrastemon*) (h)	Raffl.
Cassytha (h)	Laur.	*Notothixos*	Visc.
Cecarria	Loranth.	*Olacaceae* p.p.	Olacac.
Christisonia (h)	Orob.	*Opiliaceae* p.p.	Opil.
Cladomyza	Sant.	*Papuanthes*	Loranth.
Cuscuta (h)	Conv.	*Phacellaria*	Sant.
Cyne	Loranth.	*Rafflesia* (h)	Raffl.
Dactyliophora	Loranth.	*Rhizanthes* (h)	Raffl.
Decaisnina	Loranth.	*Rhizomonanthes*	Loranth.
Dendromyza	Sant.	*Rhopalocnemis* (h)	Balanoph.
Dendrophthoe	Loranth.	*Santalum*	Sant.
Distrianthes	Loranth.	*Scleropyrum*	Sant.
Dufrenoya	Sant.	*Scurulla*	Loranth.
Elytranthe	Loranth.	*Sogerianthe*	Loranth.
Exocarpos	Sant.	*Striga*	Scroph.
Exorhopala (h)	Balanoph.	*Taxillus*	Loranth.
Ginalloa	Visc.	*Tetradyas*	Loranth.
Helixanthera	Loranth.	*Thaumasianthes*	Loranth.
Korthalsella	Visc.	*Thesium*	Sant.
Lampas	Loranth.	*Trithecanthera*	Loranth.
Langsdorffia (h)	Balanoph.	*Viscum*	Visc.

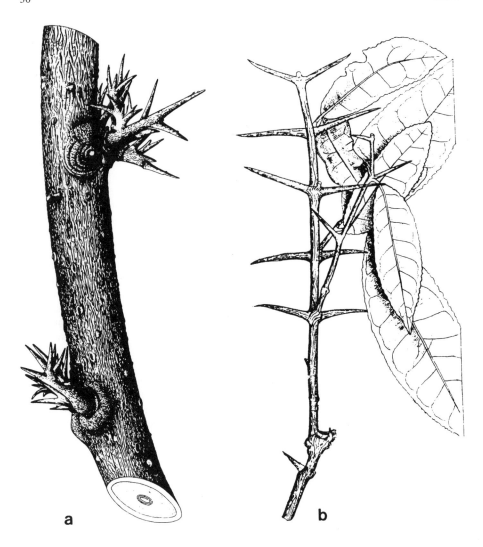

Figure 12. Armed plants – a. *Elaeagnus triflora*; b. *Paramignya longispina*.

12. Armed plants — Fig. 12 (see also Fig. 2a, p. 10)

Plants provided with thorns or spines. These can be derived from branches, stipules, or leaves. Some species are armed as juveniles and lose their thorns in the adult stage, e.g. *Cratoxylum sumatranum.*

Taxon	Family	Taxon	Family
Acacia p.p.	Leg.	*Dioscorea* p.p.	Diosc.
Acanthopanax	Aral.	*Diospyros montana*	Eben.
Acanthophora	Aral.	*Elaeagnus*	Elaeagn.
Acanthus p.p.	Acanth.	*Eleiodoxa*	Palm.
Aegle	Rut.	*Embelia* p.p.	Myrsin.
Alangium salvifolium p.p.	Alang.	*Erythrina* p.p.	Leg.
Albizia p.p.	Leg.	*Eugeissona*	Palm.
Alternanthera p.p.	Amaran.	*Euphorbia* p.p.	Euph.
Amaranthus p.p.	Amaran.	*Excoecaria indica*	Euph.
Anthorrhiza	Rub.	*Fagerlindia*	Rub.
Aralia	Aral.	*Fagraea crenulata*	Logan.
Artabotrys hexapetalus *	Annon.	*Feronia elephantum*	Rut.
Atalantia	Rut.	*Feroniella lucida*	Rut.
Azima	Salv.	*Ficus dens-echini*	Morac.
Barleria p.p.	Acanth.	*Flacourtia* p.p.	Flac.
Berberis	Berb.	*Gleditschia*	Leg.
Bombax	Bomb.	*Gmelina* p.p.	Verb.
Borassus	Palm.	*Harmsiopanax*	Aral.
Brassaiopsis	Aral.	*Harrisonia*	Simar.
Bridelia p.p.	Euph.	*Hemiscolopia*	Flac.
Bursaria *	Pitt.	*Hesperethusa crenulata*	Rut.
Cactaceae p.p.*	Cact.	*Hura* *	Euph.
Caesalpinia p.p.	Leg.	*Hydrolea spinosa*	Hydroph.
Calamus	Palm.	*Hymenocardia*	Euph.
Calospatha	Palm.	*Korthalsia*	Palm.
Canthium p.p.	Rub.	*Leea* p.p.	Leeac.
Capparis	Capp.	*Lepidium* p.p. (*Papuzilla*)	Cruc.
Cassia javanica	Leg.	*Licuala*	Palm.
Cathormion	Leg.	*Luvunga*	Rut.
Catunaregam	Rub.	*Maclura*	Morac.
Ceiba *	Bomb.	*Malpighia* *	Malp.
Ceratolobus	Palm.	*Merope*	Rut.
Ceriscoides	Rub.	*Metroxylon* p.p.	Palm.
Citriobatus	Pitt.	*Meyna*	Rub.
Citrus p.p.	Rut.	*Myrialepis*	Palm.
Cleome p.p.	Capp.	*Myrmecodia*	Rub.
Combretum quadrangulare (As)	Combr.	*Neoalsomitra* p.p.	Cuc.
Cratoxylum formosum	Gutt.	*Olax* p.p.	Olacac.
Cycas p.p.	Cycad.	*Oncosperma*	Palm.
Daemonorops	Palm.	*Oxyceros* p.p.	Rub.
Dalbergia parviflora	Leg.	*Paramignya*	Rut.
Dichrostachys p.p.	Leg.	*Parkinsonia* *	Leg.
		Pholidocarpus	Palm.

(12. Armed plants, continued)

Taxon	Family	Taxon	Family
Pigafetta	Palm.	*Scleropyrum*	Sant.
Pisonia aculeata	Nyctag.	*Scolopia*	Flac.
Planchonella punctata (As)	Sapot.	*Semecarpus bunburyanus*	Anac.
Plectocomia	Palm.	*Smilax* p.p.	Liliac.
Plectocomiopsis	Palm.	*Solanum* p.p.	Solan.
Pogonotium	Palm.	*Streblus* p.p.	Morac.
Polyscias mollis	Aral.	*Terminalia* p.p.	Combr.
Protium	Burs.	*Toddalia*	Rut.
Pterolobium	Leg.	*Trevesia*	Aral.
Punica p.p.*	Punic.	*Trifidacanthus*	Leg.
Quisqualis p.p.	Combr.	*Triphasia*	Rut.
Retispatha	Palm.	*Xanthophyllum* p.p.	Polygal.
Salacca	Palm.	*Ximenia*	Olacac.
Salsola	Chenop.	*Xylosma luzonense*	Flac.
Saurauia p.p.	Actin.	*Zanthoxylum*	Rut.
Sauropus androgynus p.p.	Euph.		

13. Bulbils

Plants provided with vegetative buds that act as diaspores. These are common in ferns but are also known in a few flowering plants. The best known example is probably *Remusatia vivipara*.

Taxon	Family	Taxon	Family
Alpinia p.p.	Zing.	*Kalanchoë*	Crass.
Caldesia	Alism.	*Nothoscordium*	Lilac.
Dioscorea p.p.	Diosc.	*Pentastemona*	Pentast.
Furcraea *	Liliac.	*Remusatia*	Arac.
Globba p.p.	Zing.	*Yucca* *	Liliac.

STEM OR BRANCH (characters 14–18)

14. Terminalia branching — Fig. 13

Branches with sympodial branching, i.e. growth at the top is arrested and elongation growth is continued from an axillary bud; well-known examples are *Baccaurea, Elaeocarpus* and *Terminalia*.

Taxon	Family	Taxon	Family
Actinodaphne	Laur.	*Palaquium*	Sapot.
Alstonia	Apoc.	*Pangium*	Flac.
Baccaurea	Euph.	*Parinari*	Chrys.
Barringtonia	Lecyth.	*Phoebe*	Laur.
Beilschmiedia	Laur.	*Pittosporum*	Pitt.
Bombax	Bomb.	*Pouteria*	Sapot.
Campnosperma	Anac.	*Rhizophoraceae* p.p.	Rhiz.
Ceiba *	Bomb.	*Rhodoleia*	Hamam.
Celtis	Ulm.	*Rubiaceae* p.p.	Rub.
Elaeocarpus	Elaeoc.	*Sapium*	Euph.
Endospermum	Euph.	*Sloanea*	Elaeoc.
Fagraea	Logan.	*Sterculia*	Sterc.
Firmiana	Sterc.	*Terminalia*	Combr.
Gluta	Anac.	*Tetractomia*	Rut.
Leguminosae p.p.	Leg.	*Theaceae* p.p.	Theac.
Manilkara	Sapot.	*Vavaea*	Meliac.

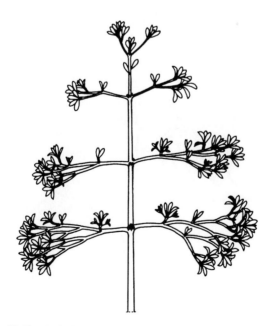

Figure 13. *Terminalia* branching. From Handb. Flora Papua New Guinea.

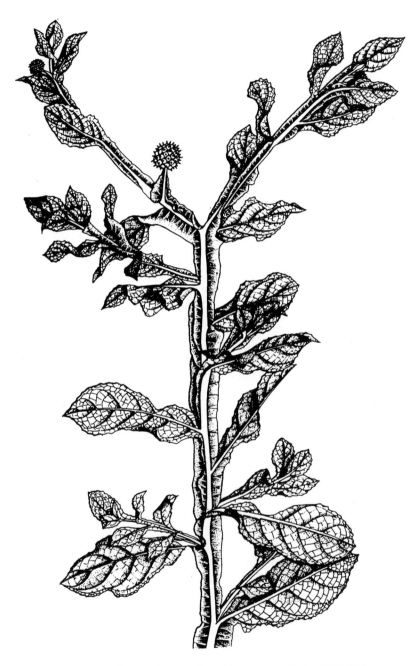

Fig. 14. Stem flanged – *Sphaeranthus africanus*. From Weeds of Rice fields.

15. Stem flanged — Fig. 14

Plants of which the stem or the branches are provided with longitudinal ridges or flanges. A good example is *Sphaeranthus*.

Taxon	Family	Taxon	Family
Alsomitra suberosa	Cuc.	*Heterostemma cuspidatum*	Asclep.
Ammobium *	Comp.	*Illigera* p.p.	Hern.
Ardisia p.p.	Myrsin.	*Laggera*	Comp.
Aristolochia crassinervia	Arist.	*Lecananthus*	Rub.
Axinandra alata	Crypter.	*Lophopetalum sessilifolium*	Celastr.
Bauhinia ridleyi	Leg.	*Medinilla* p.p.	Melast.
Cissus alata	Vit.	*Memecylon* p.p.	Melast.
Crotalaria p.p.	Leg.	*Passiflora quadrangularis* *	Passifl.
Cyrtandromoea	Scroph.	*Phyllanthus* p.p.	Euph.
Dioscorea p.p.	Diosc.	*Poikilogyne* p.p.	Melast.
Embelia p.p.	Myrsin.	*Pternandra* p.p.	Melast.
Eugenia s.l. p.p.	Myrt.	*Pterocaulon*	Comp.
Garcinia p.p.	Gutt.	*Secamone elliptica*	Asclep.
Glochidion p.p.	Euph.	*Sphaeranthus*	Comp.
Grangea	Comp.	*Strobilanthes* p.p.	Acanth.

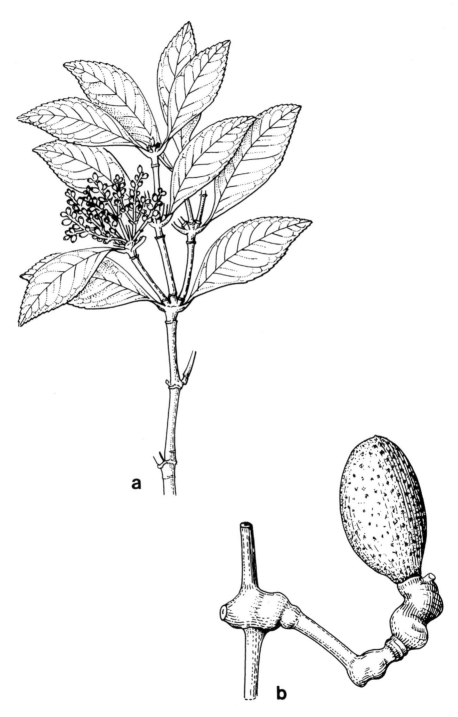

Figure 15. Swollen nodes – a. *Ascarina philippinensis*; b. *Gnetum gnemonoides*.

16. Swollen nodes — Fig. 6a, 15

Plants with branches or stems thickened at the nodes, such as in *Acanthaceae* and *Piperaceae*.

Taxon	Family	Taxon	Family
Acalypha brachystachya	Euph.	*Leea*	Leeac.
Acalypha lanceolata	Euph.	*Loranthaceae*	Loranth.
Acanthaceae	Acanth.	*Macrolenes* p.p.	Melast.
Ascarina	Chlor.	*Mallotus* p.p.	Euph.
Avicennia	Verb.	*Mirabilis* *	Nyctag.
Axinandra	Crypter.	*Peperomia*	Piper.
Carallia	Rhiz.	*Piper*	Piper.
Chloranthus	Chlor.	*Polygonum*	Polygon.
Crypteronia	Crypter.	*Pothomorphe* *	Piper.
Dactylocladus	Crypter.	*Pternandra*	Melast.
Dissochaeta p.p.	Melast.	*Sarcandra*	Chlor.
Gnetum	Gnet.	*Sonneratia*	Sonn.
Gomphrena	Amaran.	*Symingtonia*	Hamam.
Gynotroches	Rhiz.	*Thottea*	Arist.
Hedyosmum	Chlor.	*Viscaceae*	Visc.
Impatiens	Bals.	*Zippelia*	Piper.
Iresine *	Amaran.		

Figure 16. Serial buds – *Capparis zeylanica.*

17. Twigs white, petiole black

Plants when dried have pale twigs contrasting with the dark petioles, common in *Oleaceae.*

Taxon	Family	Taxon	Family
Alseodaphne p.p.	Laur.	*Ilex* p.p.	Aquif.
Beilschmiedia p.p.	Laur.	*Kayea* p.p.	Guttif.
Chionanthus p.p.	Oleac.	*Olea* p.p.	Oleac.
Corynocarpus	Coryn.	*Pittosporum* p.p.	Pitt.
Eugenia s.l., p.p.	Myrt.		

18. Serial buds — Fig. 16

Plants with several superposed buds per axil instead of one as is normal. A good example is *Capparis quiniflora.* This list is very incomplete.

Taxon	Family	Taxon	Family
Agrostistachys indica	Euph.	*Hollrungia*	Pass.
Anisophyllea	Rhiz.	*Lonicera*	Caprif.
Capparis quiniflora	Capp.	*Pithecellobium ellipticum*	Leg.
Capparis zeylanica	Capp.	*Plagiopteron* (As)	Plag.
Chisocheton p.p.	Meliac.	*Rubiaceae* p.p.	Rub.
Connarus grandis	Connar.		

EXUDATE (characters 19–22)

Many plants produce sap from wounds or cuts. Often this exudate is colourless or transparent, but in many species the exudate has a colour which may take some time to develop. Another character often invisible in the herbarium, but one that should be noted by the collector.

19. White or yellow sap

Many plants produce white (milky) sap, such as *Moraceae* and *Sapotaceae*. Yellow sap is common in *Guttiferae*. (Y) behind a name means that the sap is or can be yellow.

Taxon	Family	Taxon	Family
Aglaia	Meliac.	*Hura* *	Euph.
Anacolosa p.p.	Olacac.	*Lansium*	Meliac.
Aphanamixis	Meliac.	*Laurentia* *	Camp.
Apocynaceae	Apoc.	*Limnocharis*	Butom.
Araceae p.p.	Arac.	*Lobelia*	Lobel.
Argemone *	Papav.	*Maesa* (Y)	Myrsin.
Asclepiadaceae	Asclep.	*Mammea* (Y)	Gutt.
Burseraceae p.p.	Burs.	*Manihot* *	Euph.
Calamus (Y)	Palm.	*Mesua* (Y)	Gutt.
Calophyllum (Y)	Gutt.	*Moraceae*	Morac.
Cardiopteris	Card.	*Morinda* p.p.	Rub.
Carica *	Caric.	*Nymphaeaceae*	Nymph.
Chisocheton p.p.	Meliac.	*Ochanostachys*	Olacac.
Codiaeum	Euph.	*Omphalea*	Euph.
Codonopsis	Camp.	*Parishia* p.p.	Anac.
Compositae p.p.	Comp.	*Pimelodendron* (Y)	Euph.
Convolvulaceae p.p.	Conv.	*Pisonia umbelliflora*	Nyctag.
Cratoxylum (Y)	Gutt.	*Pleiogynium timoriense*	Anac.
Daemonorops (Y)	Palm.	*Prumnopitys ladei* (Au)	Conif.
Diploclisia (Y)	Menisp.	*Rhus* p.p.	Anac.
Elateriospermum	Euph.	*Rothmannia* p.p.	Rub.
Euphorbia	Euph.	*Salacia papuana*	Celastr.
Excoecaria	Euph.	*Sapium*	Euph.
Fagraea p.p.	Logan.	*Sapotaceae*	Sapot.
Fibraurea (Y)	Menisp.	*Stillingia*	Euph.
Ficus p.p. (Y)	Morac.	*Tenagocharis*	Butom.
Garcinia (Y)	Gutt.	*Thespesia* p.p.(Y)	Malv.
Hevea *	Euph.	*Tinomiscium* (Y)	Menisp.
Homalanthus	Euph.		

20. Black or brown sap

In most *Anacardiaceae* the sap is black or blackens upon exposure.

Taxon	Family	Taxon	Family
Anacardium *	Anac.	*Mangifera* p.p. (brown)	Anac.
Androtium	Anac.	*Melanochyla*	Anac.
Ardisia p.p.	Myrsin.	*Parishia* p.p.	Anac.
Bouea (brown)	Anac.	*Pegia*	Anac.
Buchanania	Anac.	*Pentaspadon* (brown)	Anac.
Campnosperma	Anac.	*Pistacia*	Anac.
Canarium p.p.	Burs.	*Pleiogynium*	Anac.
Dracontomelon	Anac.	*Rhus* p.p.	Anac.
Drimycarpus	Anac.	*Semecarpus*	Anac.
Euroschinus	Anac.	*Spondias*	Anac.
Gluta	Anac.	*Swintonia*	Anac.
Koordersiodendron	Anac.	*Ternstroemia* p.p.(brown)	Theac.
Lannea *	Anac.	*Triomma* p.p.	Burs.

21. Red or orange sap

Most *Myristicaceae* have red sap; in some species the sap is transparent first and only turns red after hours of exposure, in others the sap is only very faintly reddish.

Taxon	Family	Taxon	Family
Baloghia (Au P)	Euph.	*Horsfieldia*	Myrist.
Bischofia	Euph.	*Inocarpus*	Leg.
Bixa *	Bix.	*Kalappia*	Leg.
Borneodendron	Euph.	*Knema*	Myrist.
Callerya	Leg.	*Macadamia*	Prot.
Calophyllum p.p.	Gutt.	*Macaranga* p.p.	Euph.
Ceratopetalum succirubrum	Cun.	*Millettia*	Leg.
Claoxylon p.p.	Euph.	*Myristica*	Myrist.
Cochlospermum p.p.	Cochl.	*Nephelium* p.p.	Sapind.
Connarus p.p.	Connar.	*Ostodes*	Euph.
Cratoxylum p.p.	Hyper.	*Pometia pinnata*	Sapind.
Dalbergia	Leg.	*Pterocarpus* p.p.	Leg.
Dialium	Leg.	*Reutealis*	Euph.
Dysoxylum p.p.	Meliac.	*Schizomeria serrata*	Cun.
Endocomia	Myrist.	*Stephania venosa*	Menisp.
Endospermum p.p.	Euph.	*Toona sureni*	Meliac.
Fahrenheitia	Euph.	*Trigonostemon*	Euph.
Garcinia p.p.	Gutt.	*Uvaria* p.p.	Annon.
Gymnacranthera	Myrist.	*Wetria*	Euph.

22. Dried plants resinous

When being dried some plants produce a resinous substance, in a few cases so much so that the specimens stick to the newspaper in which they are dried. It is best demonstrated by *Quintinia*.

Taxon	Family	Taxon	Family
Agrostistachys	Euph.	*Owenia* (Au)	Meliac.
Anacardiaceae p.p.	Anac.	*Phaleria* p.p.	Thym.
Blumeodendron kurzii	Euph.	*Pisonia* p.p.	Nyctag.
Carallia	Rhiz.	*Pteleocarpa* p.p.	Borag.
Coffea p.p.	Rub.	*Quintinia*	Sax.
Combretocarpus	Rhiz.	*Radermachera* p.p.	Bign.
Combretum p.p.	Combr.	*Rauvolfia* p.p.	Apoc.
Dichilanthe	Rub.	*Salacia* p.p.	Celastr.
Dodonaea	Sapind.	*Sarawakodendron*	Celastr.
Elaeocarpus p.p.	Elaeoc.	*Shorea* p.p.	Dipt.
Fagraea p.p.	Logan.	*Stemonurus* p.p.	Icacin.
Garcinia p.p.	Gutt.	*Syzygium vernicosum*	Myrt.
Gardenia	Rub.	*Theaceae* p.p.	Theac.
Indorouchera	Linac.	*Trimenia*	Trim.
Lithocarpus p.p.	Fagac.	*Vandopsis lissochiloides*	Orch.
Mastixiodendron	Rub.	*Viburnum* p.p.	Caprif.
Nothofagus	Fagac.		

SMELL (characters 23, 24)

Many plants have a distinctive smell when cut or when the leaves are crushed. *Prunus* smells of almonds, *Gaultheria* of salicylic acid, *Scorodocarpus* of garlic, *Elmerillia* is fragrant, *Xanthophyllum* smells of sugarcane, etc. Most smells, however, are not distinctive and I have only taken up two categories that are recognisable even on herbarium specimens.

23. Fenugreek

Plants with a smell of fenugreek, a common ingredient of soups. I often refer to them as 'Maggi plants'. Some herbarium specimens of *Mallotus* and *Polyscias* more than 100 years old still smell of fenugreek.

Taxon	Family	Taxon	Family
Anaphalis p.p.	Comp.	*Grewia laevigata* (As)	Tiliac.
Anthoxanthum p.p.	Gram.	*Mallotus* p.p.	Euph.
Aspidopteris p.p.	Malp.	*Muehlenbergia* p.p.	Gram.
Bromheadia p.p.	Orch.	*Platea* p.p.	Icacin.
Champereia	Opil.	*Polyscias* p.p.	Aral.
Corchorus trilocularis	Tiliac.	*Ryparosa hullettii* (fr.)	Flac.
Croton p.p.	Euph.	*Sauropus* p.p.	Euph.
Cyperus hyalinus	Cyp.	*Umbelliferae* p.p.	Umb.
Eupatorium p.p.	Comp.	*Urobotrya siamensis*	Opil.

24. Foetid

Plants with a strong, disagreeable smell, often only noticeable after bruising the leaves and later disappearing in the herbarium. *Paederia foetida* is a good example. In some plants it is the flower that emits a foul smell, e.g. many *Araceae*. In that case (fl.) is added behind the name of the taxon.

Taxon	Family	Taxon	Family
Araceae p.p.(fl.)	Arac.	*Hibbertia scandens*	Dill.
Aristolochia (fl.)	Arist.	*Lasianthus* p.p.	Rub.
Cassia p.p.	Leg.	*Paederia* p.p.	Rub.
Celtis cinnamomea	Ulm.	*Polyalthia* p.p.	Annon.
Chisocheton p.p.	Meliac.	*Premna foetida*	Verb.
Cyathocalyx p.p. (fl.)	Annon.	*Rafflesiaceae* (fl.)	Raffl.
Dendrobium p.p.(fl.)	Orch.	*Saprosma*	Rub.
Dysoxylum p.p.	Meliac.	*Toona*	Meliac.
Eryngium *	Umb.		

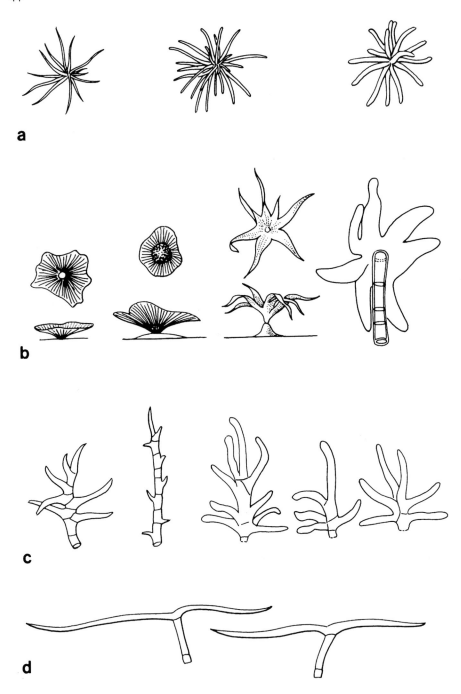

Figure 17. Indument – a. Stellate hairs; b. scales; c. dendroid hairs; d. hairs balance- or T-shaped.

INDUMENT (characters 25–30)

25. Stellate hairs — Fig. 17a

Plants of which the hairs are arranged in starshaped bundles. Very common in *Euphorbiaceae* and *Sterculiaceae*. In *Dipterocarpaceae* the hairs are often grouped in tufts. These are not regarded as true stellate hairs.

Taxon	Family	Taxon	Family
Aglaia p.p.	Meliac.	*Haplophragma adenophyllum* (As)	Bign.
Alangium p.p.	Alang.	*Heterophragma sulfureum* (As)	Bign.
Aleurites	Euph.	*Homonoia*	Euph.
Alphandia	Euph.	*Hydnocarpus* p.p.	Flac.
Aphanamixis	Meliac.	*Hydrangea*	Sax.
Astronia p.p.	Melast.	*Ipomoea* p.p.	Conv.
Baccaurea	Euph.	*Jacquemontia* p.p.	Conv.
Bombax p.p.	Bomb.	*Koilodepas*	Euph.
Borneodendron	Euph.	*Lagerstroemia*	Lythr.
Buddleja	Logan.	*Lannea* *	Anac.
Caldcluvia p.p.	Cun.	*Lunasia amara*	Rut.
Callicarpa	Verb.	*Macrolenes*	Melast.
Campnosperma p.p.	Anac.	*Mallotus* p.p.	Euph.
Capparis p.p.	Capp.	*Malvaceae* p.p.	Malv.
Catanthera	Melast.	*Medinilla*	Melast.
Cephalomappa	Euph.	*Melanolepis*	Euph.
Chisocheton p.p.	Meliac.	*Melodorum*	Annon.
Chrozophora	Euph.	*Neotrewia*	Euph.
Cladogynos	Euph.	*Olearia*	Comp.
Clethra p.p.	Clethr.	*Osmoxylon* p.p.	Aral.
Creochiton	Melast.	*Pachystylidium*	Euph.
Croton	Euph.	*Paederia foetida*	Rub.
Ctenolophon	Linac.	*Palmeria*	Monim.
Cyathocalyx p.p.	Annon.	*Pellacalyx*	Rhiz.
Dacryodes nervosa	Burs.	*Piriqueta* *	Turn.
Deutzia	Sax.	*Plagiopteron* (As)	Plag.
Dicoelia	Euph.	*Platea*	Icacin.
Diospyros p.p.	Eben.	*Rauwenhoffia*	Annon.
Dissochaeta	Melast.	*Reutealis*	Euph.
Distylium	Hamam.	*Rhododendron* p.p.	Eric.
Doryxylon	Euph.	*Rhodoleia*	Hamam.
Elaeagnus triflora	Elaeag.	*Salvia* p.p.	Lab.
Endospermum	Euph.	*Sapindaceae* p.p.	Sapind.
Epiprinus	Euph.	*Saurauia* p.p.	Actin.
Eremophila (Au)	Myopor.	*Schefflera* p.p.	Aral.
Erycibe p.p.	Conv.	*Semecarpus* p.p.	Anac.
Erythrina variegata	Leg.	*Senecio*	Comp.
Eucalyptus p.p.	Myrt.	*Shorea* p.p.	Dipt.
Fahrenheitia	Euph.	*Solanum* p.p.	Solan.
Flindersia	Rut.	*Sterculiaceae* p.p.	Sterc.
Gomphostemma	Lab.	*Styrax* p.p.	Styr.

(25. Stellate hairs, continued)

Taxon	Family	Taxon	Family
Sumbaviopsis	Euph.	*Trigonobalanus*	Fagac.
Sycopsis	Hamam.	*Uvaria*	Annon.
Tetrapanax * (As)	Aral.	*Viburnum* p.p.	Caprif.
Tiliaceae	Tiliac.	*Vitex*	Verb.
Trewia	Euph.		

26. Scales — Fig. 17b

Indument consists of round disks attached in the middle. Very common in *Durio, Elaeagnus* and *Rhododendron*.

Taxon	Family	Taxon	Family
Aglaia p.p.	Meliac.	*Heritiera*	Sterc.
Aleurites	Euph.	*Hibbertia*	Dill.
Alphitonia	Rhamn.	*Hibiscus* p.p.	Malv.
Ancistrocladus	Ancistr.	*Homonoia*	Euph.
Anisoptera p.p.	Dipt.	*Hymenocardia*	Euph.
Ardisia p.p.	Myrsin.	*Lunasia*	Rut.
Astronia p.p.	Melast.	*Macaranga* p.p.	Euph.
Berchemia	Rhamn.	*Mallotus* p.p.	Euph.
Bixa *	Bixac.	*Microcos*	Tiliac.
Brownlowia	Tiliac.	*Myrica*	Myric.
Bruinsmia	Styr.	*Myristica* p.p.	Myrist.
Callicarpa	Verb.	*Neesia*	Bomb.
Campnosperma	Anac.	*Nothofagus*	Fagac.
Camptostemon	Bomb.	*Octomeles*	Datisc.
Castanopsis	Fagac.	*Palaquium* p.p.	Sapot.
Cephalomappa	Euph.	*Parinari* p.p.	Chrys.
Chrozophora	Euph.	*Payena* p.p.	Sapot.
Chrysophyllum	Sapot.	*Pentace*	Tiliac.
Cleistanthus p.p.	Euph.	*Piriqueta* *	Turn.
Clerodendrum	Verb.	*Planchonella* p.p.	Sapot.
Coelostegia	Bomb.	*Platea*	Icacin.
Combretocarpus	Rhiz.	*Procris*	Urt.
Combretum	Combr.	*Pterospermum*	Sterc.
Croton	Euph.	*Quintinia*	Sax.
Ctenolophon	Linac.	*Raphiolepis*	Rosac.
Deutzia	Sax.	*Rhododendron*	Eric.
Diplodiscus	Tiliac.	*Schleichera*	Sapind.
Dissochaeta p.p.	Melast.	*Schoutenia*	Tiliac.
Distylium	Hamam.	*Styrax* p.p.	Styr.
Dodonaea	Sapind.	*Sumbaviopsis*	Euph.
Durio	Bomb.	*Sycopsis*	Hamam.
Elaeagnus	Elaeag.	*Thespesia*	Malv.
Engelhardia	Jugl.	*Trichospermum*	Tiliac.
Galbulimima	Himant.	*Villebrunea*	Urt.
Ganophyllum	Sapind.	*Vitex*	Verb.
Grewia	Tiliac.		

27. Dendroid hairs — Fig. 17c

Plants in which the hairs resemble miniature trees. Not a very common indument type. *Erycibe* is a genus in which this type of hair is common.

Taxon	Family	Taxon	Family
Callicarpa p.p.	Verb.	*Lagerstroemia* p.p.	Lythr.
Connarus p.p.	Connar.	*Melastoma*	Melast.
Dioscorea p.p.	Diosc.	*Myristica*	Myrist.
Erycibe p.p.	Conv.	*Platea* p.p.	Icac.
Euphorbiaceae p.p.	Euph.	*Premna* p.p.	Verb.
Indigofera	Leg.	*Scurulla* p.p.	Loranth.
Knema	Myrist.	*Vatica* p.p.	Dipt.

28. Balance hairs — Fig. 17d

Hairs not attached at base but somewhere along its length. In lateral view, these hairs are T-shaped or resemble a balance, hence the name; common in *Sapotaceae*.

Taxon	Family	Taxon	Family
Callicarpa p.p.	Verb.	*Mastixia* p.p.	Corn.
Dioscoreaceae p.p.	Diosc.	*Munronia* p.p.	Meliac.
Helicia	Prot.	*Pittosporum* p.p.	Pitt.
Icacinaceae p.p.	Icacin.	*Premna* p.p.	Verb.
Indigofera p.p.	Leg.	*Ryparosa*	Flac.
Litchi	Sapind.	*Sapotaceae* p.p.	Sapot.
Malpighiaceae p.p.	Malp.		

29. Stinging hairs

Plants provided either with sharp needle-shaped hairs that cause mechanical irritation of the skin, e.g. hairs of *Mucuna* and Bamboo, or nettle hairs such as in *Urticaceae* where irritation is mainly caused by chemical substances. In a few cases (*Brachychiton, Neesia*) stinging hairs surround the seeds; these are indicated by (fr.).

Taxon	Family	Taxon	Family
Abroma	Sterc.	*Laportea* p.p.	Urt.
Bambusoideae p.p.	Gram.	*Macaranga* p.p.	Euph.
Brachychiton (fr.)	Sterc.	*Megistostigma*	Euph.
Cnesmone	Euph.	*Mucuna*	Leg.
Dendrocnide	Urt.	*Neesia* (fr.)	Bomb.
Ficus p.p.	Morac.	*Pachystylidium*	Euph.
Fleurya p.p.	Urt.	*Phytocrene*	Icacin.
Girardinia	Urt.	*Urtica*	Urt.
Jagera	Sapind.		

30. Leaves glaucous

Many plants have leaves that are whitish or greenish underneath, caused by a waxy substance. When held against fire the wax melts. Very common in *Lauraceae*.

Taxon	Family	Taxon	Family
Alphitonia	Rhamn.	*Lauraceae* p.p.	Laur.
Anacardiaceae p.p.	Anac.	*Leguminosae* p.p.	Leg.
Annonaceae p.p.	Annon.	*Magnoliaceae* p.p.	Magn.
Daphniphyllum	Daphn.	*Menispermaceae* p.p.	Menisp.
Dipterocarpaceae p.p.	Dipt.	*Myristica* p.p.	Myrist.
Drimys p.p.	Wint.	*Rhamnus* p.p.	Rhamn.
Elaeocarpaceae p.p.	Elaeoc.	*Rubiaceae* p.p.	Rub.
Euphorbiaceae p.p.	Euph.	*Sapindaceae* p.p.	Sapind.
Eupomatia	Eupom.	*Smilax*	Liliac.
Fagaceae p.p.	Fagac.	*Trigoniastrum*	Trigon.
Hamamelidaceae p.p.	Hamam.	*Zygogynum*	Wint.
Knema	Myrist.		

LEAVES WITH GLANDS (character 31)

31. Glands on petiole (p) or lamina (l) — Fig. 18

Many plants have glands, on the petiole (p) or on the lamina (l), either on the underside or, more rarely, on the upperside. These glands are of a different type, e. g. crateriform glands of *Quassia* and *Xanthophyllum*, large flat black glands of *Prunus*, small scattered glands of *Myxopyrum*, pearl-glands of some *Macaranga* and *Flemingia*. Where no (p) or (l) is added the glands occur on both petiole and lamina or between the two.

Taxon	Family	Taxon	Family
Acacia p.p.	Leg.	*Elateriospermum* (l)	Euph.
Adenia (p)	Passifl.	*Endospermum peltatum*	Euph.
Aegialitis	Plumb.	*Eriandra* (l)	Polygal.
Ahernia (p)	Flac.	*Erythrina*	Leg.
Ailanthus (l)	Simar.	*Fagraea racemosa* (l)	Logan.
Alchornea (p)	Euph.	*Fahrenheitia* (p)	Euph.
Anacolosa p.p.	Olacac.	*Faradaya* (l)	Verb.
Ancistrocladus (l)	Ancistr.	*Fernandoa* (l)	Bign.
Anneslea (l)	Theac.	*Ficus* p.p.	Morac.
Aporosa	Euph.	*Flemingia* (l)	Leg.
Archidendron (p)	Leg.	*Gaultheria* (l)	Eric.
Ashtonia (l)	Euph.	*Gmelina* (l)	Verb.
Atylosia (l)	Leg.	*Gonystylus* (l)	Thym.
Baccaurea bracteata (l)	Euph.	*Hemiscolopia* (p)	Flac.
Bennettiodendron (p)	Flac.	*Heynea* (l)	Meliac.
Blastus (l)	Melast.	*Hibiscus* (l)	Malv.
Blumeodendron	Euph.	*Hollrungia*	Passifl.
Brucea p.p. (l)	Simar.	*Homalium*	Flac.
Bruguiera (l)	Rhiz.	*Horsfieldia* p.p. (l)	Myrist.
Cajanus (l)	Leg.	*Hosea* (l)	Verb.
Callicarpa (l)	Verb.	*Hymenocardia* (l)	Euph.
Carallia (l)	Rhiz.	*Ilex* p.p.	Aquif.
Chilocarpus (l)	Apoc.	*Itoa* (l)	Flac.
Chondrostylis (p)	Euph.	*Jasminum* p.p. (l)	Oleac.
Claoxylon	Euph.	*Koilodepas* (p)	Euph.
Clerodendrum p.p.	Verb.	*Labiatae* p.p.	Lab.
Combretocarpus	Rhiz.	*Lagenaria* (p)	Cuc.
Crateva (p)	Capp.	*Leuconotis* (l)	Apoc.
Croton	Euph.	*Ligustrum* p.p. (l)	Oleac.
Crudia p.p.	Leg.	*Limnophila* (l)	Scroph.
Deplanchea (p)	Bign.	*Lonicera* (l)	Caprif.
Desmos chinensis (p)	Annon.	*Lophopetalum* p.p. (l)	Celastr.
Dichapetalum (l)	Dichap.	*Luffa* (l)	Cuc.
Diospyros (l)	Eben.	*Macaranga* p.p.	Euph.
Diplycosia (l)	Eric.	*Macrolenes* (l)	Melast.
Dunbaria rubella (l)	Leg.	*Mallotus* p.p.	Euph.
Dysoxylum p.p. (l)	Meliac.	*Maranthes* (p)	Chrys.

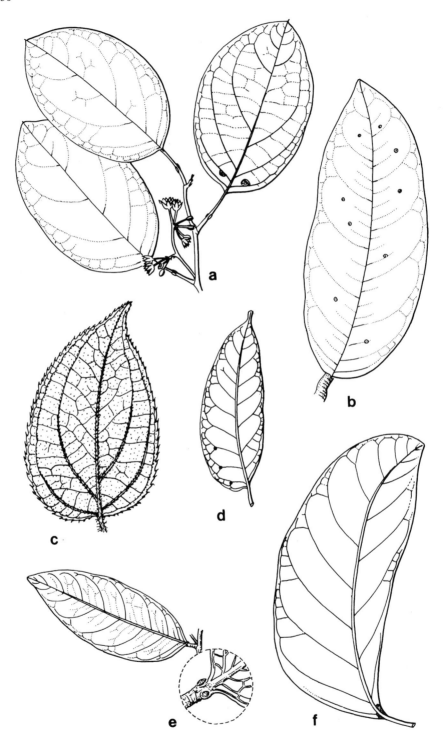

(31. Glands on petiole or lamina, continued)

Taxon	Family	Taxon	Family
Mastixia (l)	Corn.	*Rhizophora* (l)	Rhiz.
Melanolepis	Euph.	*Rhynchosia* (l)	Leg.
Momordica p.p.	Cuc.	*Rhyssopterys* (p)	Malp.
Moringa * (l)	Moring.	*Sapium* (p)	Euph.
Myxopyrum (l)	Oleac.	*Sarcosperma* p.p. (p)	Sarcosp.
Neodriessenia (l)	Melast.	*Scolopia* (p)	Flac.
Neoscortechinia (p)	Euph.	*Soulamea*	Simar.
Neosepicaea (l)	Bign.	*Stemonurus monticolus*	Icac.
Nyctocalos (l)	Bign.	*Stereospermum* (l)	Bign.
Ochanostachys p.p. (l)	Olacac.	*Stictocardia* (l)	Conv.
Octospermum (l)	Euph.	*Tecomanthe* (l)	Bign.
Pandorea (l)	Bign.	*Terminalia* p.p.	Combr.
Parastemon (p)	Chrys.	*Teijsmanniodendron* (l)	Verb.
Parinari (p)	Chrys.	*Timonius* p.p. (l)	Rub.
Paropsia	Passifl.	*Trewia* (l)	Euph.
Passiflora (p)	Passifl.	*Trichadenia*	Flac.
Perrottetia (l)	Celastr.	*Trigoniastrum* (l)	Trigon.
Pimelodendron (p)	Euph.	*Trigonostemon* (p)	Euph.
Piper p.p. (l)	Piper.	*Tristellateia* (p)	Malp.
Piriqueta *	Turn.	*Turnera* * (p)	Turn.
Polygonum p.p. (l)	Polygon.	*Vaccinium* (l)	Eric.
Polyosma (l)	Sax.	*Vatica* (l)	Dipt.
Pometia p.p. (l)	Sapind.	*Walsura* p.p. (l)	Meliac.
Prunus (l)	Rosac.	*Wetria* (l)	Euph.
Psoralea (l)	Leg.	*Xanthophyllum* (l)	Polygal.
Pullea	Cun.	*Xerospermum* (l)	Sapind.
Quassia indica (l)	Simar.	*Xylosma (p)*	Flac.
Radermachera (l)	Bign.		

←

Figure 18. Glands on petiole or lamina – a. *Hollrungia aurantioides*; b. *Quassia indica*; c. *Gaultheria abbreviata*; d. *Ailanthus triphysa*; e. *Maranthes corymbosa*; f. *Ailanthus integrifolia*.

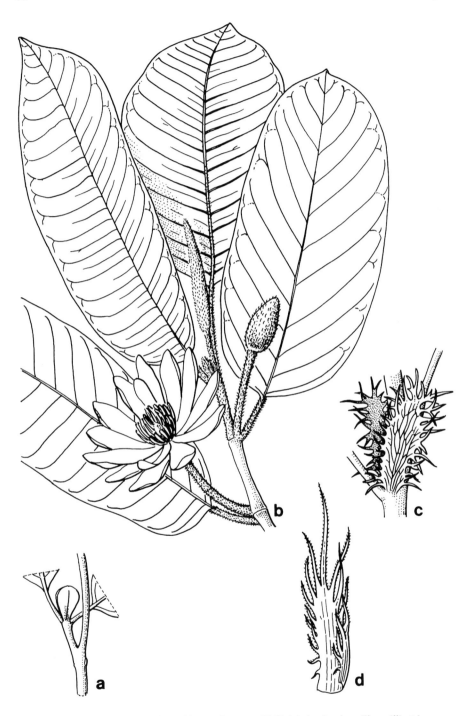

Figure 19. Stipules – a. Intrapetiolar, *Neonauclea wenzelii* (Rub.); b. clasping, *Elmerrillia tsiampacca*; c & d. pectinate, *Canarium kaniense* and *Viola pilosa*.

STIPULES (characters 32–37)

32. Intrapetiolar stipules — Fig. 19a

Plants with opposite leaves and intrapetiolar fused stipules, such as found in *Rubiaceae* and *Rhizophoraceae*. The raised ridges in some *Melastomataceae* and *Apocynaceae* are also considered intrapetiolar stipules (r).

Taxon	Family	Taxon	Family
Aizoaceae	Aizoac.	*Gynotroches*	Rhiz.
Bruguiera	Rhiz.	*Jasminum* p.p.	Oleac.
Caesalpinia oppositifolia	Leg.	*Kandelia*	Rhiz.
Callicarpa (r)	Verb.	*Lamechites* (r)	Apoc.
Carallia	Rhiz.	*Mandevilla* * (r)	Apoc.
Ceriops	Rhiz.	*Medinilla* p.p.	Melast.
Chloranthaceae (r)	Chlor.	*MicrechitesI* (r)	Apoc.
Cunoniaceae	Cun.	*Moultonianthus*	Euph.
Cynanchum p.p.	Asclep.	*Neuburgia*	Logan.
Dalenia (r)	Melast.	*Pellacalyx*	Rhiz.
Diplectria (r)	Melast.	*Rhizophora*	Rhiz.
Dissochaeta (r)	Melast.	*Rubiaceae* p.p.	Rub.
Elatine	Elat.	*Syndiophyllum*	Euph.
Erismanthus	Euph.	*Tabernaemontana* (r)	Apoc.
Fagraea	Logan.	*Turpinia*	Staph.
Ficus p.p.	Morac.		

33. Stipules clasping — Fig. 19b

Plants with spiral or alternate leaves with broadly attached stipules, leaving an annular scar, such as found in many *Moraceae* and *Magnoliaceae*.

Taxon	Family	Taxon	Family
Aegialitis	Plumb.	*Magnolia*	Magn.
Agrostistachys longifolia	Euph.	*Maingaya*	Hamam.
Agrostistachys indica	Euph.	*Manglietia*	Magn.
Araliaceae p.p.	Aral.	*Michelia*	Magn.
Artocarpus	Morac.	*Ochna*	Ochn.
Dillenia p.p.	Dill.	*Pachylarnax*	Magn.
Dipterocarpus	Dipt.	*Parashorea* p.p.	Dipt.
Elmerillia	Magn.	*Parinari* p.p.	Chrys.
Erythroxylon	Erythr.	*Piper*	Piper.
Ficus p.p.	Morac.	*Polygonaceae*	Polygon.
Gironniera	Ulm.	*Pothomorphe* *	Piper.
Gomphia	Ochn.	*Sapotaceae* p.p.	Sapot.
Houttuynia *	Saur.	*Shorea* p.p.	Dipt.
Irvingia	Simar.	*Symingtonia*	Hamam.
Leea	Leeac.	*Tadehagi*	Leg.
Macaranga p.p.	Euph.	*Zippelia*	Piper.

34. Stipules pectinate — Fig. 19c, d

Plants in which the stipules are dissected as in some *Ochnaceae*.

Taxon	Family	Taxon	Family
Acranthera	Rub.	*Koilodepas longifolia*	Euph.
Canarium p.p.	Burs.	*Koilodepas pectinata*	Euph.
Drypetes eriocarpa	Euph.	*Microcos fibrocarpa*	Tiliac.
Elaeocarpus p.p.	Elaeoc.	*Neckia*	Ochn.
Embolanthera	Hamam.	*Prunus phaeosticta*	Rosac.
Hedyotis p.p.	Rub.	*Rubus* p.p.	Rosac.
Hugonia	Linac.	*Saprosma* p.p.	Rub.
Indovethia	Ochn.	*Schuurmansia*	Ochn.
Jackiopsis	Rub.	*Viola pilosa*	Viol.

35. Stipules peltate — Fig. 20a, b

Plants in which the stipules are attached in the middle. A well-known example is *Nothofagus*.

Taxon	Family	Taxon	Family
Aeschynomene	Leg.	*Eleutherostylis*	Tiliac.
Andrachne	Euph.	*Nothofagus*	Fagac.
Aporosa p.p.	Euph.	*Phyllanthus* p.p.	Euph.
Cassia javanica	Leg.	*Prunus* p.p.	Rosac.

36. Stipules striate — Fig. 20c

Plants in which the stipules are provided with longitudinal lines or ridges. A good example is *Rinorea*

Taxon	Family	Taxon	Family
Agrostistachys p.p.	Euph.	*Hopea* p.p.	Dipt.
Bhesa	Celastr.	*Irvingia*	Simar.
Centrosema p.p.	Leg.	*Mallotus* p.p.	Euph.
Cleistanthus p.p.	Euph.	*Prismatomeris* p.p.	Rub.
Drypetes perreticulata (As)	Euph.	*Rinorea* p.p.	Viol.
Ficus p.p.	Morac.	*Shorea* p.p.	Dipt.
Gardeniopsis	Rub.	*Spatholobus* p.p.	Leg.
Heritiera p.p.	Sterc.		

Figure 20. Stipules – peltate: a. *Nothofagus nuda*; b. *Aporosa lagenocarpa*; striate: c. *Rinorea horneri*; foliaceous: d. *Canarium vulgare*; e. *Weinmannia blumei*.

37. Stipules foliaceous — Fig. 20d, e (see also Fig. 32, p. 74)

Plants with conspicuously large stipules (c. 1 cm or more across), as in many species of *Macaranga*, *Dipterocarpus*, etc.

Taxon	Family	Taxon	Family
Agrimonia	Rosac.	*Hibiscus*	Malv.
Antidesma p.p.	Euph.	*Lepisanthes* p.p.	Sapind.
Aporosa p.p.	Euph.	*Macaranga* p.p.	Euph.
Artocarpus	Morac.	*Magnoliaceae* p.p.	Magn.
Baccaurea macrophylla p.p.	Euph.	*Moultonianthus*	Euph.
Caesalpinia p.p.	Leg.	*Neillia*	Rosac.
Canarium p.p.	Burs.	*Osmelia grandistipula*	Flac.
Casearia amplectens	Flac.	*Picrasma*	Simar.
Casearia auriculata	Flac.	*Polygonum* p.p.	Polygon.
Colona	Tiliac.	*Pometia* p.p.	Sapind.
Desmodium p.p.	Leg.	*Pseudarthria*	Leg.
Dipterocarpus	Dipt.	*Rubia*	Rub.
Elaeocarpus p.p.	Elaeoc.	*Shorea* p.p.	Dipt.
Ficus	Morac.	*Sloanea*	Elaeoc.
Galium	Rub.	*Turpinia stipulacea*	Staph.
Gillbeea	Cun.	*Weinmannia*	Cun.

PETIOLE / RACHIS (charactes 38–43)

38. Petiole swollen apically — Fig. 21

Plants in which the petiole is swollen at the top and very often also at the base in which case the petiole is bipulvinate as in many *Euphorbiaceae* and *Sterculiaceae*.

Taxon	Family	Taxon	Family
Acer p.p.	Acer.	*Croton*	*Euph.*
Acronychia p.p.	Rut.	*Dapania*	Oxal.
Aglaia p.p.	Meliac.	*Deplanchea*	Bign.
Alangium p.p.	Alang.	*Desmodium* p.p.	Leg.
Albertisia	Menisp.	*Dimorphocalyx* p.p.	Euph.
Alchornea	Euph.	*Dipterocarpus*	Dipt.
Aleurites	Euph.	*Donax*	Marant.
Anamirta	Menisp.	*Durio*	Bomb.
Anisoptera	Dipt.	*Elaeocarpus* p.p.	Elaeoc.
Antidesma p.p.	Euph.	*Elateriospermum*	Euph.
Aphanamixis p.p.	Meliac.	*Eleutherandra*	Flac.
Aporosa p.p.	Euph.	*Ellipanthus*	Connar.
Araceae p.p.	Arac.	*Endospermum*	Euph.
Arcangelisia	Menisp.	*Evodia* p.p.	Rut.
Ashtonia p.p.	Euph.	*Evodiella* p.p.	Rut.
Atalantia p.p.	Rut.	*Fahrenheitia*	Euph.
Baccaurea p.p.	Euph.	*Fibraurea*	Menisp.
Baileyoxylon (Au)	Flac.	*Ficus* p.p.	Morac.
Baloghia (Au P)	Euph.	*Firmiana*	Sterc.
Bauhinia p.p.	Leg.	*Flindersia*	Rut.
Berrya	Tiliac.	*Fontainea* p.p.	Euph.
Bhesa	Celastr.	*Geijera*	Rut.
Bixa *	Bixac.	*Grewia* p.p.	Tiliac.
Blumeodendron	Euph.	*Halopegia*	Marant.
Botryophora	Euph.	*Heritiera* p.p.	Sterc.
Brachychiton	Sterc.	*Hernandia* p.p.	Hern.
Brownlowia	Tiliac.	*Heterosmilax*	Liliac.
Byttneria	Sterc.	*Hibiscus* p.p.	Malv.
Camptostemon	Bomb.	*Homalanthus* p.p.	Euph.
Carronia	Menisp.	*Hydnocarpus* p.p.	Flac.
Cephalomappa	Euph.	*Hylandia* (Au)	Euph.
Ceratopetalum	Cun.	*Hymenocardia*	Euph.
Chlaenandra	Menisp.	*Hypserpa* p.p.	Menisp.
Citrus p.p.	Rut.	*Jasminum* p.p.	Oleac.
Claoxylon p.p.	Euph.	*Lasiobema*	Leg.
Cleidion	Euph.	*Legnephora*	Menisp.
Clerodendrum schmidtii (As)	Verb.	*Limacia* p.p.	Menisp.
Codiaeum p.p.	Euph.	*Macaranga* p.p.	Euph.
Colona	Tiliac.	*Maclurodendron*	Rut.
Cominsia	Marant.	*Macrococculus*	Menisp.
Coscinium	Menisp.	*Mallotus* p.p.	Euph.

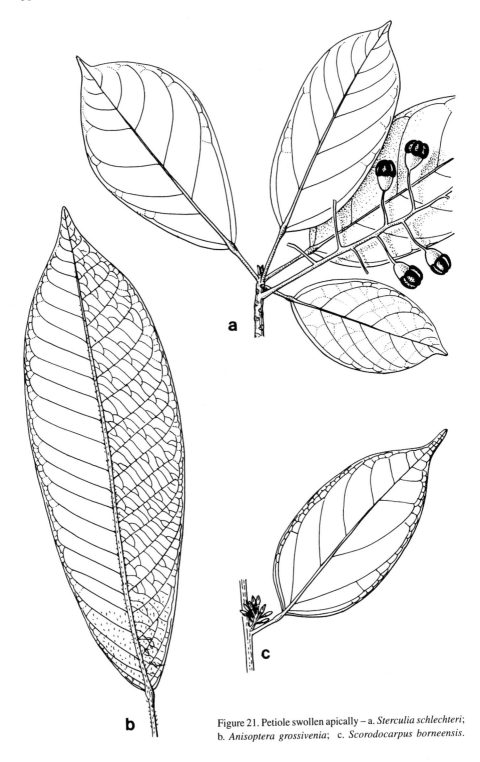

Figure 21. Petiole swollen apically – a. *Sterculia schlechteri*; b. *Anisoptera grossivenia*;   c. *Scorodocarpus borneensis*.

(38. Petiole swollen apically, continued)

Taxon	Family	Taxon	Family
Mangifera p.p.	Anac.	*Rockinghamia* (Au)	Euph.
Maranta *	Marant.	*Ryparosa*	Flac.
Maxwellia (P)	Sterc.	*Sarcopetalum*	Menisp.
Melanolepis	Euph.	*Sarcotheca* p.p.	Oxal.
Melicope p.p.	Rut.	*Scaphium*	Sterc.
Microcitrus	Rut.	*Schumannianthus*	Marant.
Millettia unifoliolata	Leg.	*Scorodocarpus*	Olac.
Monophrynium	Marant.	*Shorea* p.p.	Dipt.
Neesia	Bomb.	*Sida* p.p.	Malv.
Neoscortechinia	Euph.	*Sloanea*	Elaeoc.
Omphalea	Euph.	*Spathiostemon*	Euph.
Osmelia	Flac.	*Stachyphrynium*	Marant.
Osmoxylon	Aral.	*Stephania*	Menisp.
Pachygone	Menisp.	*Sterculia*	Sterc.
Pangium	Flac.	*Stixis*	Capp.
Paramignya	Rut.	*Streblus* p.p.	Morac.
Pentace	Tiliac.	*Teijsmanniodendron* p.p.	Verb.
Pericampylus	Menisp.	*Tetractomia* p.p.	Rut.
Phacelophrynium	Marant.	*Theobroma* *	Sterc.
Phanera	Leg.	*Thespesia*	Malv.
Phrynium	Marant.	*Thunbergia laurifolia*	Acanth.
Piliostigma	Leg.	*Tinomiscium*	Menisp.
Pimelodendron	Euph.	*Trichadenia*	Flac.
Ptychopyxis	Euph.	*Trichospermum*	Tiliac.
Pycnarrhena	Menisp.	*Upuna*	Dipt.
Reevesia	Sterc.	*Vitex* p.p.	Verb.
Rhodoleia p.p.	Hamam.	*Walsura monophylla*	Meliac.
Rhynchocarpa	Leg.	*Zanthoxylum* p.p.	Rut.

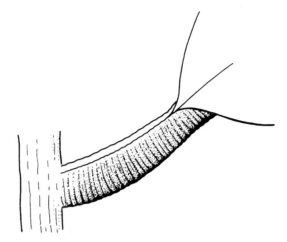

Figure 22. Petiole wrinkled – *Gonocaryum calleryanum.*

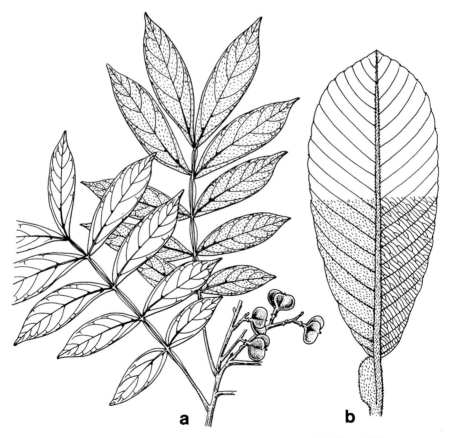

a

b

Figure 23. Winged rachis or petiole – a. *Guioa pterorhachis*; b. *Dillenia albiflos.*

39. Petiole wrinkled — Fig. 22

Petiole showing transverse ridges, very distinct in *Gonocaryum*.

Taxon	Family	Taxon	Family
Cleistanthus p.p.	Euph.	*Microtropis kinabaluensis*	Celastr.
Diospyros p.p.	Eben.	*Platea* p.p.	Icac.
Drypetes p.p.	Euph.	*Platymitra*	Annon.
Garcinia p.p.	Gutt.	*Salacia* p.p.	Celastr.
Gonocaryum	Icacin.	*Shorea* p.p.	Dipt.
Ilex p.p.	Aquif.	*Syzygium* p.p.	Myrt.
Inocarpus	Leg.	*Uvaria* p.p.	Annon.
Mammea calciphylla	Gutt.	*Xanthophyllum* p.p.	Polygal.
Mammea woodii p.p.	Gutt.		

40. Winged rachis / petiole — Fig. 23

Plants with compound leaves of which the rachis is provided with flat ridges or wings as in *Guioa* or simple leaves of which the petiole is winged as in many species of *Dillenia*.

Taxon	Family	Taxon	Family
Acrotrema	Dill.	*Inga edulis* *	Leg.
Alloxylon (*Oreocallis*)	Prot.	*Leea*	Leeac.
Archidendron pteropum	Leg.	*Lepisanthes* p.p.	Sapind.
Burkillanthus	Rut.	*Limonia*	Rut.
Campnosperma p.p.	Anac.	*Melicope* p.p.	Rut.
Citrus p.p.	Rut.	*Merrillia*	Rut.
Crescentia alata *	Bign.	*Peronema canescens*	Verb.
Davidsonia (Au)	Davids.	*Pistacia*	Anac.
Dictyoneura	Sapind.	*Pleiospermium* p.p.	Rut.
Dillenia p.p.	Dill.	*Quassia* p.p.	Simar.
Dysoxylum p.p.	Meliac.	*Sapindus*	Sapind.
Evodia p.p.	Rut.	*Tadehagi*	Leg.
Fagara *	Rut.	*Tecomanthe* p.p.	Bign.
Felicium *	Sapind.	*Teijsmanniodendron* p.p.	Verb.
Feronia elephantum	Rut.	*Toddalia* p.p.	Rut.
Grevillea p.p.	Prot.	*Turrillia*	Prot.
Guioa p.p.	Sapind.	*Vitex limonifolia*	Verb.
Harpullia	Sapind.	*Weinmannia* p.p.	Cun.
Harrisonia perforata	Simar.	*Zanthoxylum* p.p.	Rut.
Hesperethusa	Rut.		

41. Free rachis tip — Fig. 24

Compound leaves in which the rachis has a free ending, a common feature in most *Sapindaceae*.

Taxon	Family	Taxon	Family
Archidendron p.p.	Leg.	*Engelhardia* p.p.	Jugl.
Astragalus *	Leg.	*Euroschinus* p.p.	Anac.
Biophytum p.p.	Oxal.	*Parkinsonia* *	Leg.
Chisocheton p.p.	Meliac.	*Pistacia* p.p.	Anac.
Chukrasia p.p.	Meliac.	*Rutaceae* p.p.	Rut.
Dysoxylum p.p.	Meliac.	*Sapindaceae* p.p.	Sapind.

Figure 24. Free rachis tip – *Cupaniopsis stenopetala* (Sapind.).

42. Rachis with swollen nodes — Fig. 25

Compound leaves of which the rachis is swollen at the nodes. In some species, e.g. in *Oroxylum*, the rachis may break up at these nodes.

Taxon	Family	Taxon	Family
Aralia	Aral.	*Meliosma* p.p.	Sab.
Archidendron	Leg.	*Moringa* *	Moring.
Arthrophyllum	Aral.	*Oroxylum*	Bign.
Canarium p.p.	Burs.	*Picrasma*	Simar.
Dacryodes	Burs.	*Polyscias*	Aral.
Eurycoma	Simar.	*Radermachera*	Bign.
Gastonia	Aral.	*Walsura* p.p.	Meliac.
Heynea	Meliac.		
Lamiodendron	Bign.		
Leea	Leeac.		

Figure 25. Rachis with swollen nodes – *Polyscias nodosa.*

43. Petiole strongly swollen at base — Fig. 26

Plants in which the base of the petiole is conspicuously thicker than the rest of the petiole, exemplified by *Proteaceae* and *Mangifera*.

Taxon	Family	Taxon	Family
Alloxylon	Prot.	*Magnolia* p.p.	Magn.
Barringtonia	Lecyth.	*Mangifera* p.p.	Anac.
Helicia p.p.	Prot.	*Michelia* p.p.	Magn.
Heliciopsis	Prot.	*Semecarpus* p.p.	Anac.
Lithocarpus p.p.	Fagac.	*Swintonia*	Anac.
Macadamia	Prot.		

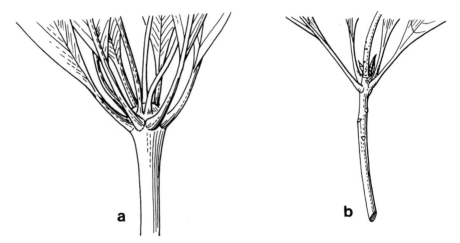

Figure 26. Petiole strongly swollen at base – a. *Macadamia hildebrandii*; b. *Barringtonia macrostachys*.

LAMINA (characters 44–69)

44. Leaves spiral in opposite-leaved families — Fig. 27

In most families the leaves are either opposite or spiral, but in some there are a few exceptions, e.g. in most *Apocynaceae* the leaves are opposite or verticillate, a few genera have spiral leaves, e.g. *Cerbera*.

Taxon	Family	Taxon	Family
Catanthera	Melast.	*Lepinia*	Apoc.
Cerbera	Apoc.	*Lepiniopsis*	Apoc.
Crescentia *	Bign.	*Medinilla* p.p.	Melast.
Dendrophthoe	Loranth.	*Melastoma* p.p.	Melast.
Gesneriaceae p.p.	Gesn.	*Plumeria* *	Apoc.
Hederella	Melast.	*Sonerila*	Melast.
Helixanthera	Loranth.	*Tristaniopsis*	Myrt.
Jasminum p.p.	Oleac.	*Xanthostemon*	Myrt.
Kjellbergiodendron	Myrt.		

Figure 27. Leaves spiral in opposite-leaved families – *Cerbera odollam.*

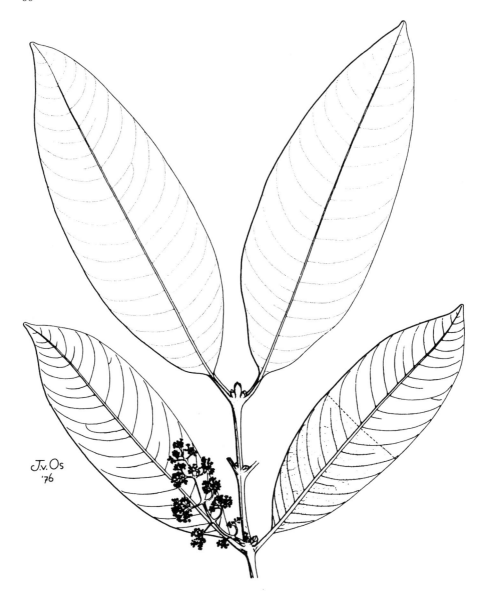

Figure 28. Leaves opposite in spiral-leaved families – *Bouea macrophylla*.

45. Leaves opposite in spiral-leaved families — Fig. 28

In most *Icacinaceae* the leaves are spiral but a few genera have opposite leaves, e.g.
Iodes. *Bouea* is the only genus of the *Anacardiaceae* with opposite leaves.

Taxon	Family	Taxon	Family
Aceratium	Elaeoc.	*Iodes*	Icacin.
Adriana (Au)	Euph.	*Litsea* p.p.	Laur.
Austrobuxus	Euph.	*Mallotus* p.p.	Euph.
Begonia p.p.	Begon.	*Moultonianthus*	Euph.
Beilschmiedia p.p.	Laur.	*Neotrewia*	Euph.
Borneodendron	Euph.	*Passiflora cochinchinensis*	Passifl.
Bouea	Anac.	*Platylobium* (Au)	Leg.
Brachysema (Au)	Leg.	*Polyporandra*	Icacin.
Caesalpinia p.p.	Leg.	*Ryparosa* p.p.	Flac.
Choriceras	Euph.	*Sapotaceae* p.p.	Sapot.
Cinnamomum p.p.	Laur.	*Saurauia* p.p.	Actin.
Citronella p.p.	Icacin.	*Scaevola* p.p.	Good.
Dysoxylum p.p.	Meliac.	*Sericolea*	Elaeoc.
Endiandra p.p.	Laur.	*Symplocos* p.p.	Sympl.
Erismanthus	Euph.	*Tournefortia* p.p.	Borag.
Excoecaria p.p.	Euph.	*Trewia*	Euph.
Ilex p.p.	Aquif.		

Figure 29. Leaves verticillate – a. *Macadamia hildebrandii*; b. *Illicium tenuifolium*.

46. Leaves verticillate — Fig. 29

More than two leaves inserted at the same level as, e.g., in *Alstonia*; when the leaves are crowded but not exactly at the same level as, e.g., in *Pittosporum* they are also considered verticillate. In this case the name is followed by (c).

Taxon	Family	Taxon	Family
Acsmithia p.p.	Cun.	*Epiprinus* (c)	Euph.
Actinodaphne	Laur.	*Eugenia* p.p.	Myrt.
Aeschynanthus p.p.	Gesn.	*Euphorbia cotinifolia* *	Euph.
Alchornea p.p. (c)	Euph.	*Faradaya* p.p.	Verb.
Allamanda *	Apoc.	*Gaertnera* p.p.	Rub.
Alseodaphne p.p.	Laur.	*Galium*	Rub.
Alseuosmia (P)	Alseu.	*Ganua pallida* (c)	Sapot.
Alstonia	Apoc.	*Garcinia* p.p.	Gutt.
Alyxia p.p.	Apoc.	*Gardenia* p.p.	Rub.
Amylotheca duthieana	Loranth.	*Geunsia* p.p.	Verb.
Angelonia *	Scroph.	*Greenea* p.p.	Rub.
Ardisia p.p.	Myrsin.	*Guettarda* p.p.	Rub.
Argostemma p.p.	Rub.	*Gymnostoma*	Casuar.
Asclepiadaceae p.p.	Asclep.	*Halfordia* p.p. (c)	Rut.
Banksia p.p.	Prot.	*Haloragis* p.p.	Halor.
Blaberopus	Apoc.	*Hamelia* p.p. *	Rub.
Blepharis p.p.	Acanth.	*Hedyotis* p.p.	Rub.
Blumeodendron p.p. (c)	Euph.	*Helicia* (c)	Prot.
Borneodendron	Euph.	*Helixanthera* p.p. (c)	Loranth.
Brasenia (c)	Nymph.	*Hydrilla*	Hydroch.
Casuarina	Casuar.	*Ilex* p.p.	Aquif.
Cephalanthus p.p.	Rub.	*Illicium* (c)	Illic.
Ceratophyllum	Cerat.	*Impatiens* p.p.	Bals.
Cerbera (c)	Apoc.	*Ixora* p.p.	Rub.
Ceuthostoma	Casuar.	*Jagera* (c)	Sapind.
Chionanthus acuminatus	Oleac.	*Kibara* p.p.	Monim.
Chloranthus henryi (As)	Chlor.	*Lampas*	Loranth.
Codiaeum p.p. (c)	Euph.	*Lantana* p.p. *	Verb.
Coelospermum p.p.	Rub.	*Lasiococca* (c)	Euph.
Coffea p.p.	Rub.	*Limnophila* p.p.	Scroph.
Combretum p.p.	Combr.	*Macadamia*	Prot.
Coprosma p.p.	Rub.	*Macaranga* p.p.	Euph.
Corynocarpus p.p. (c)	Coryn.	*Macrosolen curvinervis*	Loranth.
Crispiloba (Au)	Alseu.	*Madhuca sessilis* (c)	Sapot.
Croton p.p. (c)	Euph.	*Malpighiaceae* p.p.	Malp.
Daphniphyllum (c)	Daphn.	*Mangifera* p.p.	Anac.
Deplanchea	Bign.	*Medinilla* p.p.	Melast.
Discocalyx p.p.	Myrsin.	*Melodinus* p.p.	Apoc.
Drimys p.p. (c)	Wint.	*Mesua* p.p.	Gutt.
Dyera	Apoc.	*Meyna* p.p.	Rub.
Dysophylla	Lab.	*Mitrasacme*	Logan.
Elatine p.p.	Elat.	*Morinda* p.p.	Rub.

(46. Leaves verticillate, continued)

Taxon	Family	Taxon	Family
Mussaenda p.p.	Rub.	*Scaevola verticillata*	Good.
Myriophyllum	Halor.	*Schuurmansia* (c)	Ochn.
Myxopyrum? p.p.	Oleac.	*Scoparia* p.p.	Scroph.
Neolitsea (c)	Laur.	*Semecarpus* p.p. (c)	Anac.
Nerium *	Apoc.	*Sericolea* p.p.	Elaeoc.
Nothopegiopsis	Anac.	*Sopubia* p.p.	Scroph.
Ochrosia p.p.	Apoc.	*Sphenostemon* (c)	Sphen.
Paederia p.p.	Rub.	*Spigelia*	Logan.
Parsonsia p.p.	Apoc.	*Stemodia*	Scroph.
Pavetta p.p.	Rub.	*Swintonia* p.p. (c)	Anac.
Peperomia p.p.	Piper.	*Symplocos* p.p.	Sympl.
Pimelodendron p.p. (c)	Euph.	*Syncarpia*	Myrt.
Pisonia p.p. (c)	Nyctag.	*Terminalia* p.p. (c)	Combr.
Pittosporum p.p.	Pitt.	*Ternstroemia* (c)	Theac.
Polyosma verticillata	Sax.	*Trigonobalanus*	Fagac.
Premna p.p.	Verb.	*Trigonostemon* p.p. (c)	Euph.
Psychotria p.p.	Rub.	*Tristaniopsis* p.p. (c)	Myrt.
Pullea p.p.	Cun.	*Trithecanthera*	Loranth.
Quercus (c)	Fagac.	*Veronica* p.p.	Scroph.
Quisqualis p.p.	Combr.	*Wendlandia* p.p.	Rub.
Rauvolfia	Apoc.	*Wetria* p.p. (c)	Euph.
Rhododendron p.p. (c)	Eric.	*Wittsteinia*	Alseu.
Russelia *	Scroph.	*Wrightia* p.p.	Apoc.
Saprosma p.p.	Rub.	*Xanthostemon* p.p. (c)	Myrt.

47. Leaves anisophyllous — Fig. 30

The (members of an opposite pair of) leaves unequal in size, as in many *Acanthaceae* and *Rubiaceae* and in *Mallotus miquelianus*. Also placed in this category are some species of *Trigonostemon* with crowded leaves which are of different size.

Taxon	Family	Taxon	Family
Aeschynanthus p.p.	Gesn.	*Clerodendrum* p.p.	Verb.
Agalmyla	Gesn.	*Cypholophus nummularis*	Urt.
Aidia p.p.	Rub.	*Cyrtandra* p.p.	Gesn.
Aidiopsis p.p.	Rub.	*Cyrtandromoea*	Scroph.
Alyxia p.p.	Apoc.	*Dacrycarpus imbricatus*	Podoc.
Anerincleistus	Melast.	*Didissandra* p.p.	Gesn.
Anisophyllea	Rhiz.	*Driessenia*	Melast.
Argostemma p.p.	Rub.	*Elatostema* p.p.	Urt.
Barathranthus	Loranth.	*Geunsia* p.p.	Verb.
Blastus p.p.	Melast.	*Hallieracantha*	Acanth.
Boehmeria p.p.	Urt.	*Hedyotis* p.p.	Rub.
Callicarpa p.p.	Verb.	*Hymenodictyon*	Rub.

(30. Leaves anisophyllous, continued)

Taxon	Family	Taxon	Family
Kibara p.p.	Monim.	*Poikilogyne* p.p.	Melast.
Kochummenia p.p.	Rub.	*Porterandia* p.p.	Rub.
Leucosyke p.p.	Urt.	*Ptyssiglottis*	Acanth.
Loxonia	Gesn.	*Rhynchoglossum*	Gesn.
Lycianthes p.p.	Solan.	*Rothmannia* p.p.	Rub.
Mallotus sect. Hancea	Euph.	*Solanum* p.p.	Solan.
Maoutia p.p.	Urt.	*Sonerila*	Melast.
Medinilla p.p.	Melast.	*Stauranthera*	Acanth.
Microtoena	Lab.	*Strobilanthes* p.p.	Acanth.
Mussaenda anisophylla	Rub.	*Trewia* p.p.	Euph.
Neodriessenia	Melast.	*Tribulus*	Zygoph.
Neotrewia	Euph.	*Trigonostemon* p.p.	Euph.
Phyllagathis	Melast.		

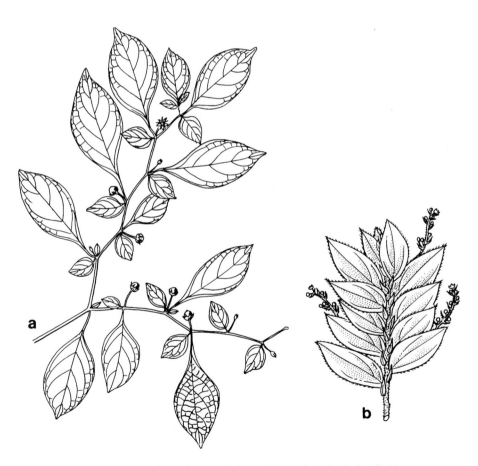

Figure 30. Leaves anisophyllous – a. *Solanum biflorum*; b. *Anisophyllea disticha.*

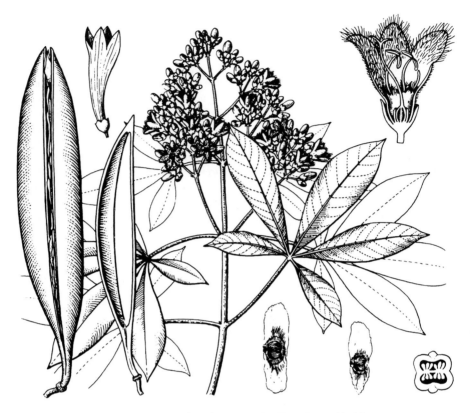

Figure 31. Leaves palmately compound – *Neosepicaea viticoides.*

48. Leaves palmately compound — Fig. 31

Leaves with three or more leaflets at the top of the petiole, e. g. most species of *Schefflera*.

Taxon	Family	Taxon	Family
Acanthopanax	Aral.	*Luvunga*	Rut.
Acronychia p.p.	Rut.	*Mackinlaya* p.p.	Aral.
Agelaea	Connar.	*Macropanax*	Aral.
Allophylus p.p.	Sapind.	*Melicope* p.p.	Rut.
Annesyoa	Euph.	*Merremia* p.p.	Conv.
Bischofia	Euph.	*Neosepicaea* p.p.	Bign.
Bombax	Bomb.	*Nyctocalos* p.p.	Bign.
Brassaiopsis	Aral.	*Osmoxylon* p.p.	Aral.
Burkillanthus p.p.	Rut.	*Oxalis*	Oxal.
Caldcluvia p.p.	Cun.	*Protium* p.p.	Burs.
Canarium p.p.	Burs.	*Rhus* p.p.	Anac.
Cannabis *	Cannab.	*Rosaceae* p.p.	Rosac.
Ceiba *	Bomb.	*Sandoricum*	Meliac.
Ceratopetalum	Cun.	*Santiria* p.p.	Burs.
Clematis p.p.	Ranunc.	*Sarcotheca* p.p.	Oxal.
Cleome	Capp.	*Schefflera* p.p.	Aral.
Connarus p.p.	Connar.	*Sterculia* p.p.	Sterc.
Crateva	Capp.	*Tecomanthe* p.p.	Bign.
Crescentia alata *	Bign.	*Teijsmanniodendron* p.p.	Verb.
Dacryodes p.p.	Burs.	*Tetractomia*	Rut.
Dioscorea p.p.	Diosc.	*Toddalia*	Rut.
Evodia p.p.	Rut.	*Trevesia*	Aral.
Geissois (P)	Cun.	*Triphasia*	Rut.
Harrisonia	Simar.	*Turpinia* p.p.	Staph.
Heritiera p.p.	Sterc.	*Vitaceae* p.p.	Vit.
Hevea *	Euph.	*Vitex* p.p.	Verb.
Illigera	Hern.	*Walsura* p.p.	Meliac.
Jasminum p.p.	Oleac.	*Weinmannia* p.p.	Cun.
Leguminosae p.p.	Leg.	*Zanthoxylum ovalifolium*	Rut.

Figure 32. Leaves compound opposite – *Gillbeea papuana*.

49. Leaves compound opposite — Fig. 32 (see also Fig. 31, p. 72)

Leaves palmately compound, pinnate or bipinnate and opposite, as, e. g., in many *Cunoniaceae* and *Bignoniaceae*.

Taxon	Family	Taxon	Family
Acronychia p.p.	Rut.	*Neosepicaea*	Bign.
Aistopetalum	Cun.	*Nyctocalos*	Bign.
Caesalpinia oppositifolia	Leg.	*Oroxylum*	Bign.
Caldcluvia p.p.	Cun.	*Pajanelia*	Bign.
Ceratopetalum p.p.	Cun.	*Pandorea*	Bign.
Clematis p.p.	Ranunc.	*Peronema*	Verb.
Compositae p.p.	Comp.	*Petraeovitex*	Verb.
Davidsonia (Au)	Davids.	*Premna* p.p.	Verb.
Dolichandrone spathacea	Bign.	*Radermachera*	Bign.
Dysoxylum p.p.	Meliac.	*Salvia scapiformis*	Lab.
Evodia p.p.	Rut.	*Sambucus*	Caprif.
Evodiella p.p.	Rut.	*Schrebera*	Oleac.
Fernandoa	Bign.	*Sesamum* *	Pedal.
Flindersia p.p.	Rut.	*Stereospermum*	Bign.
Fraxinus	Oleac.	*Tecoma* *	Bign.
Gillbeea	Cun.	*Tecomanthe*	Bign.
Hieris	Bign.	*Teijsmanniodendron* p.p.	Verb.
Jasminum p.p.	Oleac.	*Tribulus*	Zygoph.
Lamiodendron	Bign.	*Turpinia*	Staph.
Melicope p.p.	Rut.	*Valeriana*	Val.
Millingtonia	Bign.	*Vitex* p.p.	Verb.
Naravelia	Ranunc.	*Weinmannia* p.p.	Cun.

Figure 33. Leaves 2- or 3-pinnate – *Moringa oleifera.*

50. Leaves 2-, 3- (or 4-)pinnate — Fig. 33

Leaves double or triple (or quadruple) pinnate, exemplified by *Melia* and *Moringa*.

Taxon	Family	Taxon	Family
Acacia p.p.	Leg.	*Melia*	Meliac.
Acrocarpus	Leg.	*Millingtonia*	Bign.
Adenanthera	Leg.	*Mimosa* *	Leg.
Albizia	Leg.	*Moringa* *	Moring.
Ampelopsis p.p.	Vit.	*Neptunia*	Leg.
Aralia	Aral.	*Oroxylum*	Bign.
Archidendron	Leg.	*Pararchidendron*	Leg.
Archidendropsis	Leg.	*Paraserianthes*	Leg.
Artemisia p.p.	Comp.	*Parkia*	Leg.
Arthrophyllum	Aral.	*Peltophorum*	Leg.
Astilbe	Sax.	*Pithecellobium*	Leg.
Begonia bipinnatifida	Begon.	*Polyscias* p.p.	Aral.
Bidens p.p.	Comp.	*Radermachera*	Bign.
Boenninghausenia	Rut.	*Salvia* p.p.	Lab.
Caesalpinia	Leg.	*Samanea* *	Leg.
Clematis p.p.	Ranunc.	*Schefflera* p.p.	Aral.
Cosmos *	Comp.	*Schleinitzia*	Leg.
Delonix *	Leg.	*Serianthes*	Leg.
Entada	Leg.	*Stenocarpus* p.p.	Prot.
Heteropanax *	Aral.	*Thalictrum*	Ranunc.
Jacaranda *	Bign.	*Tristiropsis*	Sapind.
Leucaena *	Leg.	*Umbelliferae* p.p.	Umb.
Lomatia (Au)	Prot.	*Wallaceodendron*	Leg.

51. Leaves peltate — Fig. 34

Lamina attached away from the base as in *Nymphaea*.

Taxon	Family	Taxon	Family
Alocasia p.p.	Arac.	*Homalomena* p.p.	Arac.
Brasenia	Nymph.	*Hydrocotyle*	Umb.
Brownlowia p.p.	Tiliac.	*Hydrostemma*	Nymph.
Carica *	Caric.	*Macaranga* p.p.	Euph.
Cissampelos p.p.	Menisp.	*Mallotus* p.p.	Euph.
Colocasia p.p.	Arac.	*Megistostigma peltatum*	Euph.
Coscinium	Menisp.	*Merremia peltata*	Conv.
Cyclea	Menisp.	*Nelumbo*	Nymph.
Dendrocnide peltata	Urt.	*Nymphaea*	Nymph.
Dichondra	Conv.	*Nymphoides*	Gent.
Diploclisia p.p.	Menisp.	*Octomeles*	Datisc.
Ellipanthus beccarii var. *peltata*	Connar.	*Pothomorphe peltata* *	Piper.
Endospermum p.p.	Euph.	*Pterospermum* p.p.	Sterc.
Gonatanthus (As)	Arac.	*Sarcopetalum*	Menisp.
Harmsiopanax p.p.	Aral.	*Shorea peltata*	Dipt.
Helicia peltata	Prot.	*Stephania*	Menisp.
Hernandia p.p.	Hern.	*Tetrameles*	Datisc.
Homalanthus p.p.	Euph.		

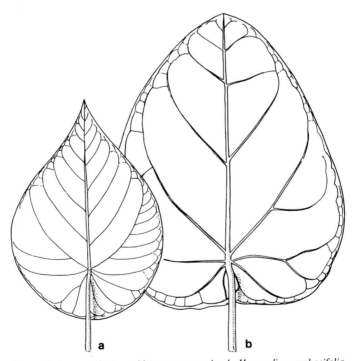

Figure 34. Leaves peltate – a. *Macaranga tanarius*; b. *Hernandia nymphaeifolia.*

52. Leaves bullate

Lamina with veins deeply sunken, so that the upper surface looks bubbly. This list is very incomplete.

Taxon	Family	Taxon	Family
Acranthera	Rub.	*Macaranga* p.p.	Euph.
Acsmithia p.p.	Cun.	*Mangifera* p.p.	Anac.
Aporosa p.p.	Euph.	*Melanochyla* p.p.	Anac.
Beilschmiedia p.p.	Laur.	*Meliosma* p.p.	Sab.
Botryophora	Euph.	*Myrica*	Myric.
Caldcluvia brassii	Cun.	*Olearia* p.p.	Comp.
Carpodetus	Sax.	*Oreomitra* p.p.	Cuc.
Cypholophus chamaephyton	Urt.	*Oxyspora* p.p.	Melast.
Cyrtandra p.p.	Gesn.	*Poikilogyne villosa*	Melast.
Didissandra p.p.	Gesn.	*Polyalthia* p.p.	Annon.
Didymocarpus p.p.	Gesn.	*Pullea* p.p.	Cun.
Dioscorea p.p.	Diosc.	*Quercus* p.p.	Fagac.
Diospyros p.p.	Eben.	*Rhyticaryum*	Icacin.
Elaeocarpus p.p.	Elaeoc.	*Sericolea* p.p.	Elaeoc.
Ficus p.p.	Morac.	*Shorea* p.p.	Dipt.
Gonystylus areolatus	Thym.	*Symplocos* p.p.	Sympl.
Helicia bullata	Prot.	*Syzygium* p.p.	Myrt.
Hydrostemma	Nymph.	*Tetractomia* p.p.	Rut.
Ilex p.p.	Aquif.	*Urophyllum* p.p.	Rub.
Kayea p.p.	Gutt.	*Weinmannia* p.p.	Cun.
Koilodepas p.p.	Euph.	*Willughbeia anomala*	Apoc.
Laportea decumana	Urt.	*Xanthomyrtus* p.p.	Myrt.
Lasianthus p.p.	Rub.	*Zygogynum* p.p.	Wint.
Lonicera p.p.	Caprif.		

53. Dicots with large leaves — Fig. 35

Adult leaves more than 40 cm long or across as in several species of *Campnosperma* and *Dillenia*.

Taxon	Family	Taxon	Family
Agrostistachys	Euph.	*Meryta* p.p. (P)	Aral.
Anakasia	Aral.	*Monophyllaea* p.p.	Gesn.
Antidesma p.p.	Euph.	*Neesia* p.p.	Bomb.
Artocarpus p.p.	Morac.	*Octamyrtus* (Polak 1134, 1256)	Myrt.
Aulandra	Sapot.	*Osmoxylon* p.p.	Aral.
Barringtonia p.p.	Lecyth.	*Parashorea* p.p.	Dipt.
Brassaiopsis p.p.	Aral.	*Piper* p.p.	Piper.
Campnosperma p.p.	Anac.	*Poikilospermum* p.p.	Urt.
Carica *	Caric.	*Polyalthia dolichophylla*	Annon.
Codiaeum (Avé 4740)	Euph.	*Pothomorphe* *	Piper.
Dillenia p.p.	Dill.	*Pterospermum*	Sterc.
Diplodiscus p.p.	Tiliac.	*Saurauia* p.p.	Actin.
Dipterocarpus p.p.	Dipt.	*Scaphium*	Sterc.
Dolicholobium p.p.	Rub.	*Schuurmansia*	Ochn.
Drypetes p.p.	Euph.	*Semecarpus* p.p.	Anac.
Elaeocarpus gustaviifolius	Elaeoc.	*Shorea* p.p.	Dipt.
Ficus p.p.	Morac.	*Sterculia* p.p.	Sterc.
Garcinia p.p.	Gutt.	*Streptocarpus*	Gesn.
Goniothalamus p.p.	Annon.	*Tapeinosperma* p.p.	Myrsin.
Gonystylus areolatus	Thym.	*Tectona* p.p.	Verb.
Harmsiopanax	Aral.	*Trevesia* p.p.	Aral.
Macaranga p.p.	Euph.	*Trigonostemon sandakanensis*	Euph.
Magnolia p.p.	Magn.	*Vatica* p.p.	Dipt.
Mallotus p.p.*?*	Euph.	*Wetria macrophylla*	Euph.
Mammea p.p.	Gutt.	*Xanthophyllum adenotus*	Polygal.

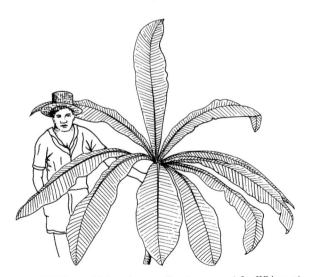

Figure 35. Dicots with large leaves – *Tapeinosperma* (after Whitmore).

54. Nigrescence

Leaves turning blackish upon drying as in many *Rubiaceae, Diospyros* etc.

Taxon	Family	Taxon	Family
Acranthera p.p.	Rub.	*Morinda*	Rub.
Annonaceae p.p.	Annon.	*Mucuna*	Leg.
Apodytes	Icacin.	*Myrmecodia*	Rub.
Aralidium	Aral.	*Pavetta* p.p.	Rub.
Argostemma p.p.	Rub.	*Pilea* p.p.	Urt.
Breynia	Euph.	*Pisonia*	Nyctag.
Buchnera	Scroph.	*Platanthera*	Orch.
Calanthe	Orch.	*Porterandia* p.p.	Rub.
Canthium p.p.	Rub.	*Psychotria* p.p.	Rub.
Celtis	Ulm.	*Rothmannia* p.p.	Rub.
Cerbera	Apoc.	*Santalum*	Sant.
Coprosma	Rub.	*Saprosma*	Rub.
Dehaasia	Laur.	*Scaevola*	Good.
Dendromyza	Sant.	*Scyphostegia*	Scyph.
Diospyros p.p.	Eben.	*Striga*	Scroph.
Dolichandrone	Bign.	*Strychnos*	Logan.
Geniostoma	Logan.	*Tarenna* p.p.	Rub.
Gynochthodes	Rub.	*Tournefortia*	Borag.
Heliotropium	Borag.	*Urophyllum nigricans*	Rub.
Hydnophytum	Rub.	*Viscum*	Visc.
Ixora p.p.	Rub.	*Vitex negundo*	Verb.
Mastersia	Leg.	*Voacanga*	Apoc.
Messerschmidia	Borag.	*Ximenia*	Olacac.

55. Dry leaves yellow

Leaves turning yellow upon drying; very common in *Symplocos* and *Xanthophyllum*.

Taxon	Family	Taxon	Family
Actephila p.p.	Euph.	*Glycosmis* p.p.	Rut.
Alangium p.p.	Alang.	*Helicia* p.p.	Prot.
Anisophyllea p.p.	Rhiz.	*Lindsayomyrtus*	Myrt.
Aporosa frutescens	Euph.	*Memecylon* p.p.	Melast.
Ashtonia	Euph.	*Polyosma* p.p.	Sax.
Baccaurea p.p.	Euph.	*Rinorea* p.p.	Viol.
Ceratopetalum	Cun.	*Symplocos* p.p.	Sympl.
Diospyros toposia	Eben.	*Syzygium* p.p.	Myrt.
Elaeocarpus p.p.	Elaeoc.	*Xanthophyllum* p.p.	Polygal.
Ficus diversifolia	Morac.		

56. Young leaves red

In many plants the juvenile leaves are red. Unfortunately this is not always mentioned on the labels.

Taxon	Family	Taxon	Family
Acer	Acer.	*Guttiferae* p.p.	Gutt.
Anacardiaceae p.p.	Anac.	*Lauraceae* p.p.	Laur.
Annonaceae p.p.	Annon.	*Leguminosae* p.p.	Leg.
Connaraceae	Connar.	*Myrtaceae* p.p.	Myrt.
Dipterocarpaceae p.p.	Dipt.	*Sapindaceae* p.p.	Sapind.
Ericaceae p.p.	Eric.	*Theaceae* p.p.	Theac.
Euphorbiaceae p.p.	Euph.		

57. Broken leaves with white threads

This feature is best seen in fresh material. When a leaf is broken the two parts adhere to each other by white threads, the spiral rings of the tracheids or the dried contents of resin ducts.

Taxon	Family	Taxon	Family
Aleurites	Euph.	*Hevea* *	Euph.
Annonaceae p.p.	Annon.	*Lauraceae* p.p.	Laur.
Apocynaceae p.p.	Apoc.	*Linostoma*	Thym.
Aquilaria	Thym.	*Loranthaceae* p.p.	Loranth.
Croton	Euph.	*Macaranga*	Euph.
Euonymus	Celastr.	*Mangifera* p.p.	Anac.
Eurycoma	Simar.	*Moraceae*	Morac.
Excoecaria	Euph.	*Ochanostachys*	Olacac.
Fahrenheitia	Euph.	*Sapotaceae*	Sapot.
Gnetum	Gnet.		

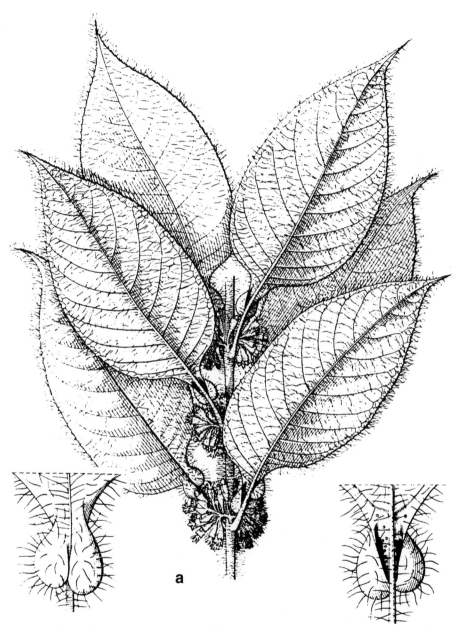

Figure 36. Leaves with domatia – a. *Callicarpa saccata* (Verb.). See also next page.

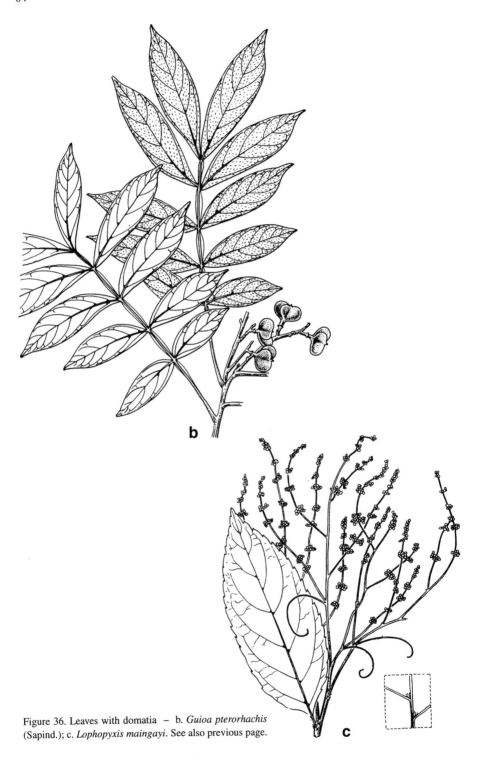

Figure 36. Leaves with domatia – b. *Guioa pterorhachis*
(Sapind.); c. *Lophopyxis maingayi*. See also previous page.

58. Leaves with domatia — Fig. 36

Leaves with hairy or membranous structures on the underside, in the axils of nerves, often inhabited by mites or ants.

Taxon	Family	Taxon	Family
Acer p.p.	Acer.	*Magnoliaceae* p.p.	Magn.
Alangium	Alang.	*Malvaceae* p.p.	Malv.
Anacardiaceae p.p.	Anac.	*Mastixia*	Corn.
Annonaceae p.p.	Annon.	*Melastomataceae* p.p.	Melast.
Apocynaceae p.p.	Apoc.	*Meliaceae* p.p.	Meliac.
Araliaceae p.p.	Aral.	*Menispermaceae* p.p.	Menisp.
Bignoniaceae p.p.	Bign.	*Moraceae* p.p.	Morac.
Boraginaceae p.p.	Borag.	*Myrtaceae* p.p.	Myrt.
Burseraceae p.p.	Burs.	*Nyssa*	Nyss.
Caprifoliaceae p.p.	Caprif.	*Olacaceae* p.p.	Olacac.
Celastraceae p.p.	Celastr.	*Oleaceae* p.p.	Oleac.
Clethra	Clethr.	*Piperaceae* p.p.	Piper.
Cochlospermum	Cochl.	*Polygalaceae* p.p.	Polygal.
Combretaceae p.p.	Combr.	*Rhamnaceae* p.p.	Rhamn.
Compositae p.p.	Comp.	*Rosaceae* p.p.	Rosac.
Cunoniaceae p.p.	Cun.	*Rubiaceae* p.p.	Rub.
Dilleniaceae p.p.	Dill.	*Rutaceae* p.p.	Rut.
Dipterocarpaceae p.p.	Dipt.	*Sapindaceae* p.p.	Sapind.
Elaeocarpaceae p.p.	Elaeoc.	*Sarcosperma*	Sarcosp.
Engelhardia	Jugl.	*Scrophulariaceae* p.p.	Scroph.
Euphorbiaceae p.p.	Euph.	*Simaroubaceae* p.p.	Simar.
Fagaceae p.p.	Fagac.	*Solanaceae* p.p.	Solan.
Flacourtiaceae p.p.	Flac.	*Sterculiaceae* p.p.	Sterc.
Gesneriaceae p.p.	Gesn.	*Styracaceae* p.p.	Styr.
Hamamelidaceae p.p.	Hamam.	*Theaceae* p.p.	Theac.
Hernandiaceae p.p.	Hern.	*Tiliaceae* p.p.	Tiliac.
Icacinaceae p.p.	Icacin.	*Ulmaceae* p.p.	Ulm.
Ilex p.p.	Aquif.	*Urticaceae* p.p.	Urt.
Lauraceae p.p.	Laur.	*Verbenaceae* p.p.	Verb.
Lophopyxis	Loph.	*Violaceae* p.p.	Viol.
Lythraceae p.p.	Lythr.	*Vitaceae* p.p.	Vit.

59. Leaves with dots — Fig. 37

The dots become visible when the leaf is held against strong light (use handlens!).
They appear as small transparent (*Rutaceae*) or coloured (*Myrsinaceae*) dots.

Taxon	Family	Taxon	Family
Acanthaceae p.p.	Acanth.	*Caesalpinia*	Leg.
Aetoxylon	Thym.	*Callicarpa*	Verb.
Aglaia p.p.	Meliac.	*Cansjera*	Opil.
Amyxa	Thym.	*Capparis*	Capp.
Anacolosa	Olacac.	*Carnarvonia* (Au)	Prot.
Annonaceae p.p.	Annon.	*Casearia* p.p.	Flac.
Anogeissus (As)	Combr.	*Celtis*	Ulm.
Astronia	Melast.	*Chionanthus* p.p.	Oleac.
Buckinghamia (Au)	Prot.	*Cissus*	Vit.
Buxus p.p.	Bux.	*Colona*	Tiliac.

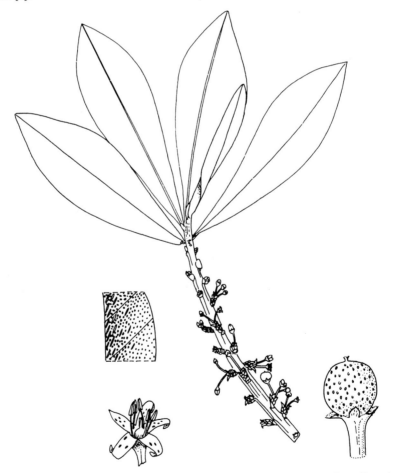

Figure 37. Leaves with dots – *Rapanea involucrata* (Myrsin.) (courtesy Dr. P. van Royen).

(59. Leaves with dots, continued)

Taxon	Family	Taxon	Family
Combretum	Combr.	*Mischocarpus*	Sapind.
Compositae p.p.	Comp.	*Monimiaceae*	Monim.
Connarus	Connar.	*Morus*	Morac.
Cordia p.p.	Borag.	*Myoporum*	Myopor.
Corynocarpus	Coryn.	*Myristica* p.p.	Myrist.
Cratoxylum	Gutt.	*Myrsinaceae* (not *Maesa*)	Myrsin.
Crotalaria	Leg.	*Myrtaceae*	Myrt.
Croton	Euph.	*Octomeles* p.p.	Datisc.
Dendropanax borneensis	Aral.	*Osmelia* p.p.	Flac.
Derris thyrsiflora	Leg.	*Peperomia*	Piper.
Diospyros p.p.	Eben.	*Piper* p.p.	Piper.
Dodonaea	Sapind.	*Podocarpus* p.p.	Conif.
Drimys	Wint.	*Polyosma*	Sax.
Elaeocarpus p.p.	Elaeoc.	*Pothomorphe* *	Piper.
Ficus p.p.	Morac.	*Prunus* p.p.	Rosac.
Fontainea	Euph.	*Psoralea*	Leg.
Galbulimima	Himant.	*Rutaceae*	Rut.
Garcinia p.p.	Gutt.	*Ryparosa*	Flac.
Geissois (P)	Cun.	*Salacia macrophylla*	Celastr.
Glochidion	Euph.	*Santalaceae* p.p.	Sant.
Gnetum	Gnet.	*Sarcopteryx*	Sapind.
Gonystylus	Thym.	*Schisandra*	Schis.
Guioa	Sapind.	*Scorodocarpus*	Olacac.
Heynea p.p.	Meliac.	*Smilax*	Liliac.
Hypericum	Gutt.	*Stixis*	Capp.
Icacinaceae p.p.	Icacin.	*Suregada*	Euph.
Illicium	Illic.	*Sympetalandra*	Leg.
Illigera	Hern.	*Syndiophyllum*	Euph.
Jacquemontia	Conv.	*Terminalia* p.p.	Combr.
Kadsura	Schis.	*Ternstroemia* p.p.	Theac.
Kingiodendron	Leg.	*Tetrastigma*	Vit.
Labiatae	Lab.	*Timonius* p.p.	Rub.
Lauraceae p.p.	Laur.	*Trimenia*	Trim.
Lysimachia p.p.	Prim.	*Urticaceae*	Urt.
Mammea	Gutt.	*Vitex* p.p.	Verb.
Memecylon p.p.	Melast.	*Walsura* p.p.	Meliac.
Merrilliodendron	Icacin.	*Zygogynum*	Wint.

60. Leaf surface puncticulate

Leaf with tiny depressions as if pricked with a needle.

Taxon	Family	Taxon	Family
Acanthus ilicifolius	Acanth.	*Eugenia* p.p.	Myrt.
Agelaea	Connar.	*Hosea*	Verb.
Amyxa	Thym.	*Macaranga* p.p.	Euph.
Anneslea	Theac.	*Prunus*	Rosac.
Aphanamyxis polystachya	Meliac.	*Rhododendron* p.p.	Eric.
Aporosa p.p.	Euph.	*Sarcosperma*	Sarcosp.
Avicennia	Verb.	*Teijsmanniodendron*	Verb.
Chionanthus p.p.	Oleac.	*Trimenia macrura*	Trim.
Dichapetalum	Dichap.	*Viburnum punctatum*	Caprif.

61. Leaf surface pustulate

Leaf surface with small raised swellings, often giving the lamina a dull appearance; common in *Loranthaceae* and *Olacaceae*.

Taxon	Family	Taxon	Family
Euphorbiaceae p.p.	Euph.	*Loranthaceae* p.p.	Loranth.
Fagraea p.p.	Logan.	*Memecylon* p.p.	Melast.
Flacourtiaceae p.p.	Flac.	*Olacaceae*	Olacac.
Horsfieldia p.p.	Myrist.	*Opiliaceae*	Opil.
Icacinaceae p.p.	Icacin.	*Popowia*	Annon.
Jasminum p.p.	Oleac.		

62. Leaf surface rough

Leaf surface very rough to the touch, hence the term 'sandpaper' leaves; common in several species of *Ficus* and *Tetracera*.

Taxon	Family	Taxon	Family
Artocarpus p.p.	Morac.	*Horsfieldia grandis*	Myrist.
Broussonetia *	Morac.	*Hullettia*	Morac.
Claoxylon p.p.	Euph.	*Macaranga trachyphylla*	Euph.
Cucurbitaceae p.p.	Cuc.	*Tetracera* p.p.	Dill.
Didymocarpus p.p.	Gesn.	*Trema cannabina*	Ulm.
Dillenia pentagyna	Dill.	*Urticaceae* p.p.	Urt.
Ficus p.p.	Morac.	*Wedelia asperrima*	Comp.
Homalomena asperifolia	Arac.		

63. Cystoliths

Leaves provided with cells containing silica crystals, visible as raised pale dots or dashes (use handlens).

Taxon	Family	Taxon	Family
Acanthaceae p.p.	Acanth.	*Hedyotis*	Rub.
Amaracarpus	Rub.	*Mallotus* p.p.	Euph.
Argostemma	Rub.	*Melastoma* p.p.	Melast.
Arisaema	Arac.	*Moraceae* p.p.	Morac.
Astronia p.p.	Melast.	*Mycetia*	Rub.
Astronidium p.p.	Melast.	*Nertera*	Rub.
Baliospermum	Euph.	*Piper* p.p.	Piper.
Bougainvillea *	Nyctag.	*Rhaphidophora* p.p.	Arac.
Cordia	Borag.	*Saurauia*	Actin.
Cryptocoryne	Arac.	*Urticaceae*	Urt.
Dioscorea p.p.	Diosc.		

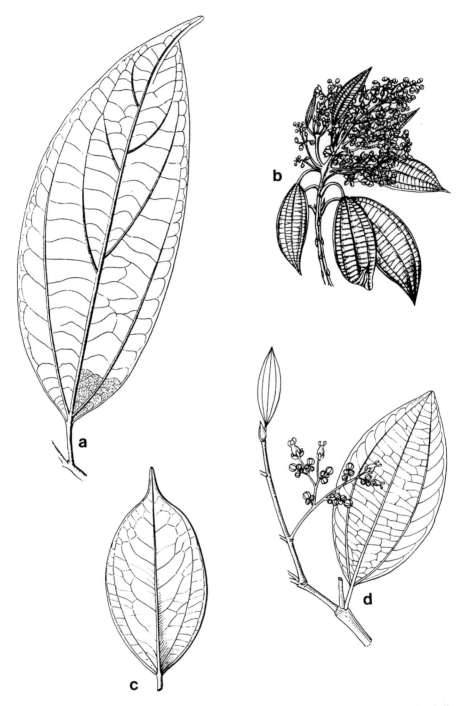

Figure 38. Leaves triplinerved – a. *Cryptocarya densiflora*; b. *Astronia spectabilis* (Melast.); c. *Anisophyllea disticha*; d. *Celtis philippensis*.

64. Leaves triplinerved — Fig. 38

Leaves with a pair of opposite veins at the base, which may reach the top of the lamina (e.g. *Cinnamomum*) or end somewhere in the leaf margin (e.g. *Ficus*).

Taxon	Family	Taxon	Family
Adenia	Passifl.	*Gomphandra quadrifida*	
Alangium p.p.	Alang.	var. *triplinervis*	Icacin.
Amyema p.p.	Loranth.	*Grewia* p.p.	Tiliac.
Anisophyllea	Rhiz.	*Jasminum* p.p.	Oleac.
Austromuellera (Au)	Prot.	*Leptonychia*	Sterc.
Berrya	Tiliac.	*Leucosyke*	Urt.
Blumeodendron p.p.	Euph.	*Lindera* p.p.	Laur.
Boehmeria	Urt.	*Macaranga* p.p.	Euph.
Brachychiton p.p.	Sterc.	*Mallotus* p.p.	Euph.
Brackenridgea	Ochn.	*Maoutia*	Urt.
Brownlowia	Tiliac.	*Melastomataceae* p.p.	Melast.
Callitriche	Callitr.	*Microcos*	Tiliac.
Caryodaphnopsis	Laur.	*Myxopyrum*	Oleac.
Celtis	Ulm.	*Neolitsea* p.p.	Laur.
Cinnamomum p.p.	Laur.	*Notothixos*	Visc.
Clematis	Ranunc.	*Osmelia*	Flac.
Cocculus laurifolius	Menisp.	*Palmeria*	Monim.
Colona	Tiliac.	*Pentace* p.p.	Tiliac.
Colubrina anomala	Rhamn.	*Piper*	Piper.
Commersonia	Sterc.	*Pipturus*	Urt.
Coriaria	Coriar.	*Pouzolzia*	Urt.
Crawfurdia	Gent.	*Rhodamnia*	Myrt.
Cryptocarya p.p.	Laur.	*Rhodomyrtus*	Myrt.
Cucurbitaceae p.p.	Cuc.	*Ryparosa*	Flac.
Debregeasia	Urt.	*Salomonia*	Polygal.
Dendrocnide p.p.	Urt.	*Sarcococca*	Bux.
Dioscorea	Diosc.	*Schoepfia*	Oleac.
Diplodiscus	Tiliac.	*Schoutenia*	Tiliac.
Diplycosia	Eric.	*Scolopia* p.p.	Flac.
Disporum	Liliac.	*Scorodocarpus*	Olacac.
Ericaceae p.p.	Eric.	*Smilax*	Liliac.
Erythropalum	Olacac.	*Stemona*	Stem.
Erythrospermum	Flac.	*Sterculia*	Sterc.
Exacum	Gent.	*Strychnos*	Logan.
Faradaya	Verb.	*Thottea*	Arist.
Fatoua	Morac.	*Thunbergia*	Acanth.
Ficus p.p.	Morac.	*Trema*	Ulm.
Galium	Rub.	*Trewia*	Euph.
Gaultheria	Eric.	*Trichospermum*	Tiliac.
Gentiana	Gent.	*Trigonostemon* p.p.	Euph.
Gibbsia	Urt.	*Villebrunea*	Urt.
Ginalloa	Visc.	*Viscum*	Visc.
		Zizyphus	Rhamn.

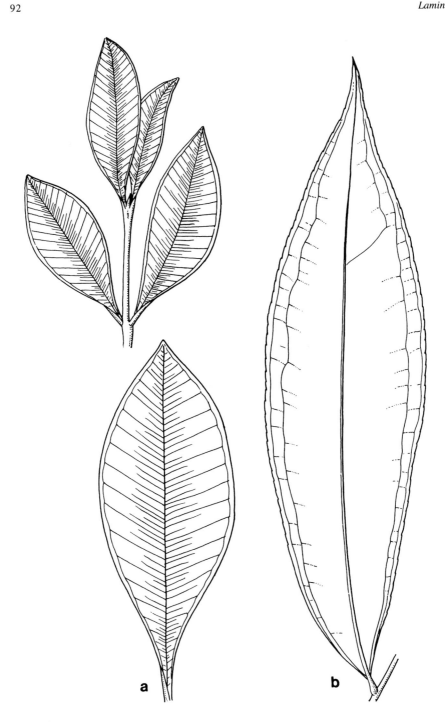

Figure 39. a. Intramarginal vein – *Eugenia suringariana* (Myrt.). — b. Double intramarginal vein – *Gomphia serrata*.

65. Intramarginal vein — Fig. 39a

A vein running parallel to the margin of the lamina (e.g. *Eugenia s. l.*). The distinction between a triplinerved leaf, a leaf with intramarginal vein and one in which the veins are looped and joined is not always easy to observe.

Taxon	Family	Taxon	Family
Anisophyllea	Rhiz.	*Leuconotis*	Apoc.
Bridelia	Euph.	*Memecylon* p.p.	Melast.
Buxaceae	Bux.	*Monocarpia*	Annon.
Chilocarpus	Apoc.	*Myrtaceae*	Myrt.
Crypteroniaceae	Crypter.	*Sapotaceae* p.p.	Sapot.
Drimycarpus	Anac.	*Scaphocalyx*	Flac.
Duabanga	Sonn.	*Spondias* p.p.	Anac.
Finschia	Prot.	*Swintonia*	Anac.
Gomphia	Ochn.		

66. Double intramarginal vein — Fig. 39b

Two veins running parallel to the leaf margin, e.g. in *Gomphia*.

Taxon	Family	Taxon	Family
Axinandra	Crypter.	*Octamyrtus* p.p.	Myrt.
Buxus p.p.	Bux.	*Pedicellarum*	Arac.
Decaspermum p.p.	Myrt.	*Pothos*	Arac.
Gomphia	Ochn.	*Syzygium* p.p.	Myrt.
Nepenthes	Nepenth.	*Whiteodendron*	Myrt.

67. Parallel secondary venation — Fig. 40

Leaves with very close parallel veins; *Calophyllum* is the best known example.

Taxon	Family	Taxon	Family
Alstonia	Apoc.	*Linostoma*	Thym.
Amyxa	Thym.	*Mimusops* p.p.	Sapot.
Aquilaria	Thym.	*Musa*	Musac.
Calophyllum	Gutt.	*Neckia*	Ochn.
Carallia caryophylloidea	Rhiz.	*Palaquium* p.p.	Sapot.
Chrysophyllum p.p.	Sapot.	*Payena* p.p.	Sapot.
Dryobalanops	Dipt.	*Reinwardtiodendron humile*	Meliac.
Euthemis	Ochn.	*Schuurmansia*	Ochn.
Ficus p.p.	Morac.	*Schuurmansiella*	Ochn.
Garcinia p.p.	Gutt.	*Sericolea* p.p.	Elaeoc.
Gonystylus	Thym.	*Severinia* p.p.	Rut.
Gyrinops caudata	Thym.	*Tephrosia* p.p.	Leg.
Hopea p.p.	Dipt.	*Timonius* p.p.	Rub.
Indovethia	Ochn.	*Wetria*	Euph.
Kayea calophylloides	Gutt.		

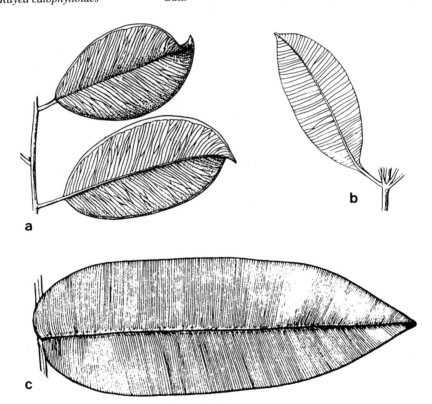

Figure 40. Parallel secondary venation – a. *Gonystylus bancanus*; b. *Alstonia angustiloba*; c. *Calophyllum complanatum*.

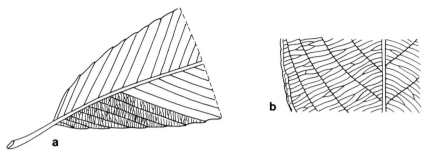

Figure 41. Scalariform venation – a. *Dillenia indica*; b. *Rinorea horneri*.

68. Scalariform venation — Fig. 41

The tertiary veins are close and parallel (ladder-like), common in *Rhamnaceae*.

Taxon	Family	Taxon	Family
Alphitonia	Rhamn.	*Macaranga*	Euph.
Antidesma p.p.	Euph.	*Maesopsis* *	Rhamn.
Aporosa p.p.	Euph.	*Mallotus*	Euph.
Atuna	Chrys.	*Maranthes*	Chrys.
Baccaurea p.p.	Euph.	*Melastomataceae* p.p.	Melast.
Berchemia	Rhamn.	*Neobalanocarpus*	Dipt.
Bhesa	Celastr.	*Parashorea*	Dipt.
Bridelia	Euph.	*Parinari*	Chrys.
Colubrina	Rhamn.	*Rhamnella*	Rhamn.
Combretaceae p.p.	Combr.	*Rhamnus*	Rhamn.
Desmodium p.p.	Leg.	*Rinorea* p.p.	Viol.
Dilleniaceae p.p.	Dill.	*Sageretia*	Rhamn.
Dipterocarpus	Dipt.	*Sapotaceae* p.p.	Sapot.
Emmenosperma	Rhamn.	*Scorodocarpus*	Olacac.
Enkleia	Thym.	*Shorea* p.p.	Dipt.
Flemingia	Leg.	*Smythea*	Rhamn.
Gouania	Rhamn.	*Stemona*	Stem.
Grewia	Tiliac.	*Upuna*	Dipt.
Hopea p.p.	Dipt.	*Ventilago*	Rhamn.
Irvingia	Simar.	*Zizyphus*	Rhamn.
Lasianthus	Rub.		

69. Leaves withering red

This is again a feature not visible in the herbarium and one depends on completeness of the label. A good example is provided by *Elaeocarpus*.

Taxon	Family	Taxon	Family
Acer	Acer.	*Lagerstroemia*	Lythr.
Elaeocarpus	Elaeoc.	*Sapium*	Euph.
Greenea	Rub.	*Terminalia*	Combr.
Homalanthus	Euph.	*Wendlandia*	Rub.

INFLORESCENCE (characters 70–78)

70. Cauliflorous plants — Fig. 42

Plants with the inflorescences borne on the stem or trunk. This is not always clear in a
herbarium specimen. So, the condition should be stated on the label.

Taxon	Family	Taxon	Family
Actinorhytis	Palm.	*Aulandra*	Sapot.
Aglaia p.p.	Meliac.	*Averrhoa* p.p.	Oxal.
Alchornea borneensis	Euph.	*Baccaurea* p.p.	Euph.
Anacolosa cauliflora	Olacac.	*Barringtonia* p.p.	Lecyth.
Anamirta	Menisp.	*Bellucia* *	Melast.
Annona p.p.*	Annon.	*Callerya*	Leg.
Antidesma p.p.	Euph.	*Callicarpa* p.p.	Verb.
Arcangelisia	Menisp.	*Caryota*	Palm.
Archidendron p.p.	Leg.	*Chisocheton* p.p.	Meliac.
Areca	Palm.	*Chlaenandra*	Menisp.
Arenga	Palm.	*Coscinium* p.p.	Menisp.
Artabotrys p.p.	Annon.	*Couroupita* *	Lecyth.
Artocarpus p.p.	Morac.	*Cyathocalyx biovulatus*	Annon.

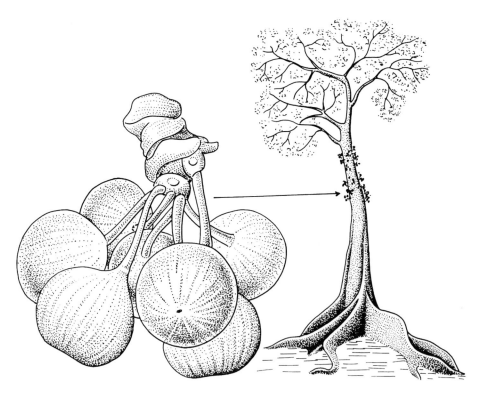

Figure 42. Cauliflorous plants – *Ficus variegata.*

(70. *Cauliflorous plants, continued*)

Taxon	Family	Taxon	Family
Cyathostemma	Annon.	*Palaquium beccarii*	Sapot.
Cyclea	Menisp.	*Pandanus* p.p.	Pand.
Cynometra cauliflora	Leg.	*Parmentiera* *	Bign.
Cyrtandra p.p.	Gesn.	*Phaleria* p.p.	Thym.
Cyrtostachys	Palm.	*Phyllanthus acidus*	Euph.
Diospyros p.p.	Eben.	*Phytocrene* p.p.	Icacin.
Diploclisia	Menisp.	*Pimelodendron macrocarpum*	Euph.
Drypetes p.p.	Euph.	*Pinanga*	Palm.
Durio p.p.	Bomb.	*Pisonia* p.p.	Nyctag.
Dysoxylum p.p.	Meliac.	*Planchonella keyensis* p.p.	Sapot.
Enicosanthum p.p.	Annon.	*Polyalthia* p.p.	Annon.
Eugenia p.p.	Myrt.	*Praravinia suberosa*	Rub.
Evodia p.p.	Rut.	*Premna* p.p.	Verb.
Faradaya p.p.	Oleac.	*Pseudobotrys*	Icacin.
Ficus p.p.	Morac.	*Ptychopyxis grandiflorus*	Euph.
Fordia p.p.	Leg.	*Quassia* p.p.	Simar.
Forrestia	Comm.	*Radermachera* p.p.	Bign.
Galearia celebica p.p.	Euph.	*Rhopaloblaste*	Palm.
Glochidion p.p.	Euph.	*Rhynchotechum*	Gesn.
Gnetum p.p.	Gnet.	*Ryparosa* p.p.	Flac.
Goniothalamus p.p.	Annon.	*Saraca*	Leg.
Gonocaryum p.p.	Icacin.	*Sarcopetalum*	Menisp.
Haematocarpus p.p.	Menisp.	*Saurauia* p.p.	Actin.
Helicia p.p.	Prot.	*Sauropus* p.p.	Euph.
Heliciopsis p.p.	Prot.	*Scaphocalyx*	Flac.
Illicium p.p.	Illic.	*Schefflera* p.p. (*Burley 3346*)	Aral.
Ixora p.p.	Rub.	*Scleropyrum*	Sant.
Kadsura	Schis.	*Semecarpus* p.p.	Anac.
Lansium p.p.	Meliac.	*Steganthera* p.p.	Monim.
Lepisanthes p.p.	Sapind.	*Stelechocarpus*	Annon.
Litsea p.p.	Laur.	*Stephania*	Menisp.
Lycianthes p.p.	Solan.	*Sterculia* p.p.	Sterc.
Macrococculus	Menisp.	*Stichianthus*	Rub.
Magodendron	Sapot.	*Strongylodon* p.p.	Leg.
Mammea woodii	Gutt.	*Tetrastigma*	Vit.
Mayodendron igneum (As)	Bign.	*Theobroma* *	Sterc.
Melientha	Opil.	*Tiliacora*	Menisp.
Merrilliodendron	Icacin.	*Tinomiscium*	Menisp.
Moultonia	Gesn.	*Trigonostemon capillipes*	Euph.
Mucuna p.p.	Leg.	*Urophyllum* p.p.	Rub.
Nenga	Palm.	*Uvaria*	Annon.
Octamyrtus p.p.	Myrt.	*Versteeghia*	Rub.
Oncosperma	Palm.	*Wallichia*	Palm.
Opuntia *	Cact		

Figure 43. Inflorescence fasciculate, leaves distichous – a. *Scorodocarpus borneensis*; b. *Rinorea horneri*; c. *Lindera lucida*.

71. Inflorescence fasciculate, leaves distichous — Fig. 43

This combination is characteristic for many genera in various families, e.g. *Euphorbiaceae* and *Flacourtiaceae*.

Taxon	Family	Taxon	Family
Actephila p.p.	Euph.	*Knema*	Myrist.
Anacolosa p.p.	Olacac.	*Leptopus*	Euph.
Aporosa p.p.	Euph.	*Lindera* p.p.	Laur.
Boehmeria p.p.	Urt.	*Litsea* p.p.	Laur.
Breynia	Euph.	*Margaritaria*	Euph.
Bridelia	Euph.	*Microdesmis*	Euph.
Casearia	Flac.	*Myristica* p.p.	Myrist.
Chaetocarpus	Euph.	*Paropsia* p.p.	Passifl.
Chamabainia	Urt.	*Phyllanthus*	Euph.
Cleistanthus p.p.	Euph.	*Pouzolzia*	Urt.
Cypholophus	Urt.	*Procris* p.p.	Urt.
Dichapetalum p.p.	Dichap.	*Rapanea*	Myrsin.
Diospyros p.p.	Eben.	*Rinorea* p.p.	Viol.
Drypetes p.p.	Euph.	*Sapotaceae* p.p.	Sapot.
Elatostema	Urt.	*Sauropus*	Euph.
Ellipanthus p.p.	Connar.	*Scolopia* p.p.	Flac.
Flueggea	Euph.	*Scorodocarpus*	Olac.
Glochidion p.p.	Euph.	*Sebastiania* p.p.	Euph.
Gonostegia	Urt.	*Strombosia*	Olacac.
Gyrinops p.p.	Thym.	*Suregada*	Euph.
Hemiscolopia	Flac.	*Trigonopleura*	Euph.
Hydnocarpus p.p.	Flac.	*Trigonostemon* p.p.	Euph.
Kairothamnus	Euph.	*Ximenia* p.p.	Olacac.

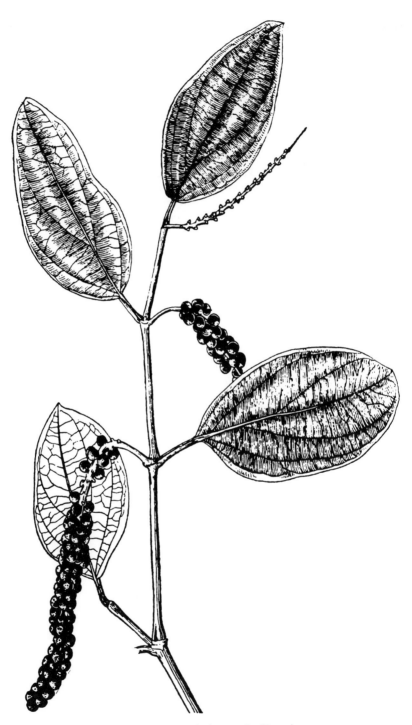

Figure 44. Inflorescence leaf-opposed – *Piper nigrum.*

72. Inflorescence leaf-opposed — Fig. 44

Inflorescence borne opposite the leaf instead of in the leaf axil. Well known examples are *Piper* and *Suregada*.

Taxon	Family	Taxon	Family
Abroma p.p.	Sterc.	*Lycianthes* p.p.	Solan.
Allmannia	Amaran.	*Macaranga* p.p.	Euph.
Ampelocissus	Vit.	*Magnoliaceae* p.p.	Magn.
Anaxagorea p.p.	Annon.	*Mallotus* p.p.	Euph.
Aporosa	Euph.	*Monocarpia* p.p.	Annon.
Cissus	Vit.	*Pelargonium* *	Geran.
Commersonia	Sterc.	*Peperomia* p.p.	Piper.
Commelinaceae p.p.	Comm.	*Piper* p.p.	Piper.
Cyathocalyx p.p.	Annon.	*Plukenetia* p.p.	Euph.
Fissistigma p.p.	Annon.	*Solanum* p.p.	Solan.
Gomphandra p.p.	Icacin.	*Spathiostemon* p.p.	Euph.
Houttuynia p.p. *	Saur.	*Suregada*	Euph.
Leguminosae p.p.	Leg.	*Zippelia* p.p.	Piper.
Lepiniopsis	Apoc.		

73. Inflorescence supra-axillary — See Fig. 16, p. 38

Inflorescence (or flower) not in the leaf axil but above it, e.g. *Glyptopetalum*.

Taxon	Family	Taxon	Family
Aidia	Rub.	*Gaertnera* p.p.	Rub.
Annonaceae p.p.	Annon.	*Gardenia* p.p.	Rub.
Capparis p.p.	Capp.	*Glyptopetalum*	Celastr.
Chionanthus	Oleac.	*Hydnocarpus* p.p.	Flac.
Citronella p.p.	Icacin.	*Neckia*	Ochn.
Cowiea	Rub.	*Oleaceae* p.p.	Oleac.
Diospyros p.p.	Eben.	*Polygala* p.p.	Polygal.
Fordia	Leg.	*Stichianthus*	Rub.

74. Inflorescence epiphyllous — Fig. 45

Stalk of inflorescence (or flower) fused with leaf. Very rare in Malesia. A good example is *Ruthiella*.

Taxon	Family
Chisocheton p.p.	Meliac.
Didissandra morganii	Gesn.
Helwingia *	Corn.
Monophyllaea	Gesn.
Neuropeltopsis	Conv.
Ruthiella	Camp.
Solanum p.p.	Solan.
Trianthema portulacastrum	Aizoac.
Turnera *	Turn.

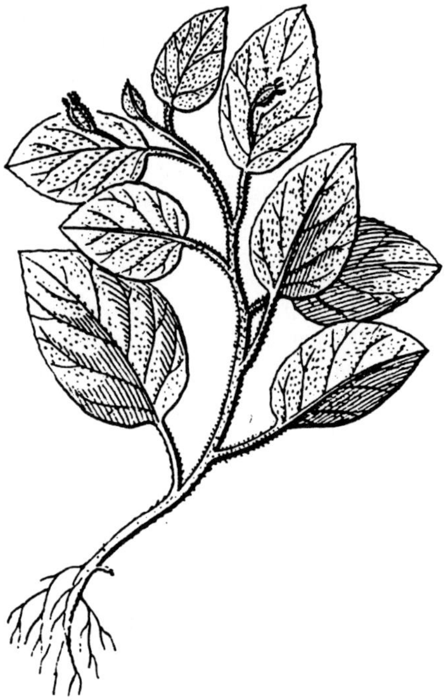

Figure 45. Inflorescence epiphyllous – *Ruthiella nigrum.*

75. Geocarpous plants — Fig. 46

Inflorescence subterranean, as in some species of *Ficus*, or originally above ground, entering the soil later as in *Arachis*.

Taxon	Family	Taxon	Family
Arachis *	Leg.	*Goniothalamus* p.p.	Annon.
Artocarpus p.p.?	Morac.	*Neocolletia*	Leg.
Commelina benghalensis	Comm.	*Saurauia* p.p.	Actin.
Cyrtandra p.p.	Gesn.	*Uvaria* p.p.	Annon.
Desmos p.p.	Annon.	*Vigna* p.p.	Leg.
Enicosanthum p.p.	Annon.	*Voandzeia* *	Leg.
Ficus p.p.	Morac.	*Zingiberaceae* p.p.	Zing.

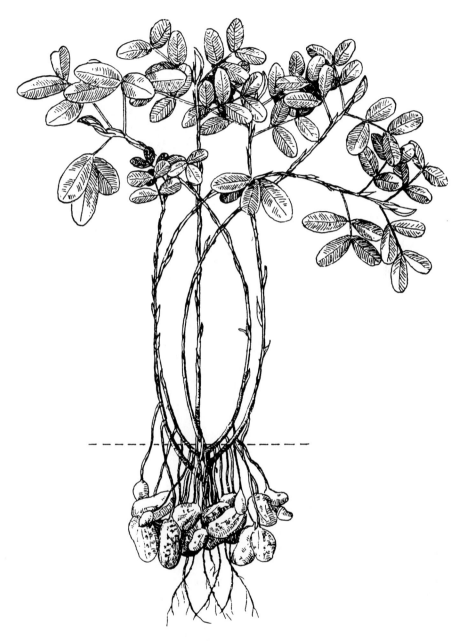

Figure 46. Geocarpous plants – *Arachis hypogaea.*

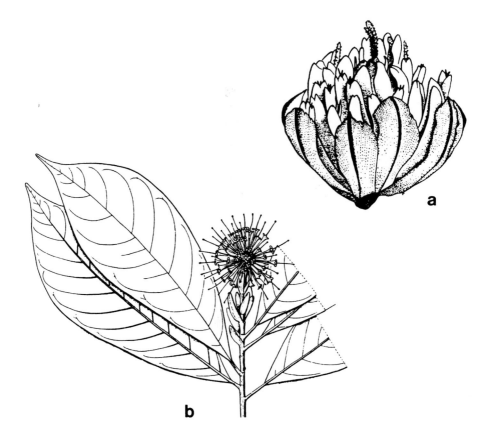

Figure 47. Inflorescence compact – a. *Sphaeranthus africanus* (Comp.); b. *Myrmeconauclea stipulacea* (Rub.).

76. Inflorescence compact — Fig. 47

Flowers sitting tightly together in a head as in *Compositae, Uncaria* etc.

Taxon	Family	Taxon	Family
Actinodaphne	Laur.	*Cephalomappa*	Euph.
Altingia	Hamam.	*Ceuthostoma*	Casuar.
Anakasia	Aral.	*Cladogynos*	Euph.
Annanas *	Brom.	*Compositae*	Comp.
Anogeissus (As)	Combr.	*Coniferae* p.p.	Conif.
Araceae	Arac.	*Cyperaceae* p.p.	Cyp.
Astrothalamus	Urt.	*Daphne*	Thym.
Caldcluvia p.p.	Cun.	*Epiprinus*	Euph.
Casuarina	Casuar.	*Eriocaulon*	Erioc.
Celosia	Amaran.	*Eryngium* *	Umb.

(76. Inflorescence compact, continued)

Taxon	Family	Taxon	Family
Freycinetia	Pand.	*Pimelea*	Thym.
Gomphrena	Amaran.	*Pterisanthes*	Vit.
Gramineae p.p.	Gram.	*Ptilotus*	Amaran.
Gymnostoma	Casuar.	*Pullea p.p*	Cun.
Koilodepas	Euph.	*Rhodoleia*	Hamam.
Leguminosae p.p.	Leg.	*Rubiaceae* p.p.	Rub.
Lepeostegeres	Loranth.	*Sararanga*	Pand.
Lepidaria	Loranth.	*Saurauia* p.p.	Actin.
Lindera	Laur.	*Schefflera* p.p.	Aral.
Litsea	Laur.	*Scyphostegia*	Scyph.
Meryta (P)	Aral.	*Sparganium*	Sparg.
Moraceae	Morac.	*Symingtonia*	Hamam.
Myrtaceae p.p.	Myrt.	*Typha*	Typh.
Nypa	Palm.	*Urticaceae* p.p.	Urt.
Nyssa	Nyss.	*Xyris*	Xyr.
Pandanus	Pand.		

77. Inflorescence a condensed raceme — Fig. 48

Basically a raceme but flowers very close together as in *Kopsia* and *Scyphostegia*.

Taxon	Family
Embelia p.p.	Myrsin.
Euphorbiaceae p.p.	Euph.
Hoya p.p.	Asclep.
Kopsia p.p.	Apoc.
Rapanea p.p.	Myrsin.
Rubiaceae p.p.	Rub.
Sarawakodendron	Celastr.
Scyphostegia	Scyph.

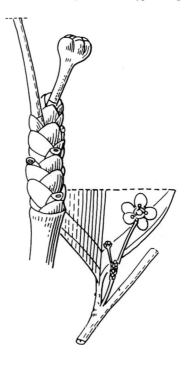

Figure 48. Inflorescence a condensed raceme –
Sarawakodendron filamentosum.

Figure 49. Flagelliflory – a. *Barringtonia scortechinii*; b. *Quassia indica* (→).

78. Flagelliflory — Fig. 49

Inflorescence long and pendent, usually terminal, e.g. *Barringtonia* and *Parkia*.

Taxon	Family	Taxon	Family
Aglaia p.p.	Meliac.	*Galearia*	Euph.
Alpinia p.p.	Zing.	*Ixora* p.p.	Rub.
Antidesma p.p.	Euph.	*Kigelia* *	Bign.
Aphanamixis	Meliac.	*Macadamia hildebrandii*	Prot.
Arenga	Palm.	*Meliosma* p.p.	Sab.
Baccaurea p.p.	Euph.	*Mucuna*	Leg.
Barringtonia	Lecyth.	*Musa* p.p.	Musac.
Calamus p.p.	Palm.	*Octomeles*	Datisc.
Calyptrocalyx	Palm.	*Parkia*	Leg.
Carronia	Menisp.	*Petalophus*	Annon.
Chisocheton p.p.	Meliac.	*Piper* p.p.	Piper.
Cowiea	Rub.	*Plectocomia*	Palm.
Dendrocnide	Urt.	*Quassia indica*	Simar.
Diospyros p.p.	Eben.	*Strongylodon*	Leg.
Diploclisia	Menisp.	*Tinomiscium*	Menisp.
Engelhardia	Jugl.	*Toona*	Meliac.
Eurycoma	Simar.		
Fahrenheitia p.p.	Euph.		
Fibraurea	Menisp.		

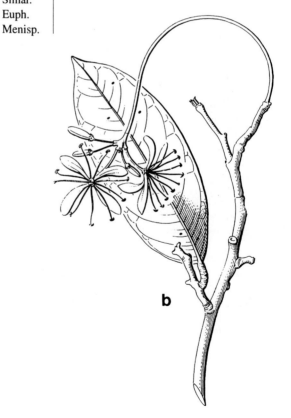

b

FLOWER (characters 79–92)

79. 3-merous dicots

Most dicots are 5-merous, 3-merous flowers are a rule in some dicot families such as *Annonaceae, Lauraceae* and *Menispermaceae*.

Taxon	Family	Taxon	Family
Ailanthus p.p.	Simar.	*Hernandiaceae* p.p.	Hern.
Anisophyllea p.p.	Rhiz.	*Icacinaceae* p.p.	Icacin.
Annonaceae p.p.	Annon.	*Kandelia* p.p.	Rhiz.
Araliaceae p.p.	Aral.	*Lauraceae*	Laur.
Aristolochiaceae p.p.	Arist.	*Loranthaceae* p.p.	Loranth.
Balanophora p.p.	Balanoph.	*Magnoliaceae* p.p.	Magn.
Bennettiodendron p.p.	Flac.	*Malaisia*	Morac.
Berberidaceae p.p.	Berb.	*Menispermaceae*	Menisp.
Bouea p.p.	Anac.	*Myristicaceae*	Myrist.
Buxaceae p.p.	Bux.	*Olacaceae* p.p.	Olacac.
Campnosperma p.p.	Anac.	*Onagraceae* p.p.	Onagr.
Canarium p.p.	Burs.	*Palaquium*	Sapot.
Ceriops p.p.	Rhiz.	*Piperaceae* p.p.	Piper.
Cheilotheca p.p.	Eric.	*Polygonaceae* p.p.	Polygon.
Combretocarpus p.p.	Rhiz.	*Quassia* p.p.	Simar.
Cunoniaceae p.p.	Cun.	*Ranunculaceae* p.p.	Ranunc.
Dacryodes p.p.	Burs.	*Salacia* p.p.	Celastr.
Daphniphyllum p.p.	Daphn.	*Santiria* p.p.	Burs.
Diospyros p.p.	Eben.	*Santalaceae* p.p.	Sant.
Elatine p.p.	Elat.	*Saururus* p.p.	Saur.
Euphorbiaceae p.p.	Euph.	*Scolopia* p.p.	Flac.
Eurycoma p.p.	Simar.	*Scyphostegia* p.p.	Scyph.
Fagaceae p.p.	Fagac.	*Sonerila*	Melast.
Guttiferae p.p.	Gutt.	*Soulamea* p.p.	Simar.
Haplolobus p.p.	Burs.	*Tetracera* p.p.	Dill.
Hemiscolopia p.p.	Flac.	*Winteraceae* p.p.	Wint.

80. Calyx accrescent — Fig. 50

Calyx increasing in size after anthesis, as, e.g., in *Diospyros* and many *Dipterocarpaceae*.

Taxon	Family	Taxon	Family
Actephila p.p.	Euph.	*Clerodendrum*	Verb.
Ancistrocladus	Ancistr.	*Dimorphocalyx*	Euph.
Antigonon *	Polygon.	*Diospyros*	Eben.
Ardisia p.p.	Myrsin.	*Dipterocarpaceae*	Dipt.
Blachia	Euph.	*Drypetes* p.p.	Euph.
Breynia	Euph.	*Epiprinus*	Euph.
Capparis p.p.	Capp.	*Erismanthus* p.p.	Euph.

(80. Calyx accrescent, continued)

Taxon	Family	Taxon	Family
Faradaya	Verb.	*Parishia*	Anac.
Garcinia	Gutt.	*Petraeovitex*	Verb.
Gluta p.p.	Anac.	*Physalis*	Solan.
Harmandia	Olacac.	*Schoutenia*	Tiliac.
Hernandia	Hern.	*Swintonia*	Anac.
Holmskioldia *	Verb.	*Theaceae* p.p.	Theac.
Hymenopyramis (As)	Verb.	*Trigonostemon*	Euph.
Koilodepas pectinata	Euph.	*Vitex*	Verb.
Lasiococca	Euph.		

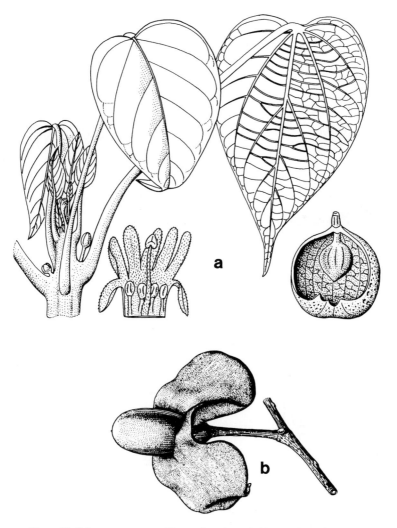

Figure 50. Calyx accrescent – a. *Hernandia ovigera*; b. *Harmandia mekongensis*.

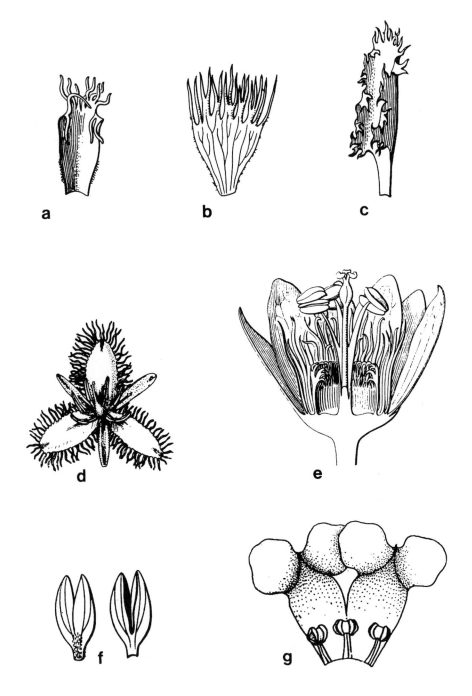

Figure 51. Corolla / petals fimbriate / bifid – a. *Ceriops tagal*; b. *Elaeocarpus stipularis*; c. *Carallia brachiata*; d. *Thysanotus tuberosus*; e. *Hollrungia aurantioides* (Passifl.); f. *Dichapetalum timoriense*; g. *Erycibe griffithii*.

81. Corolla / petals fimbriate / bifid — Fig. 51

Plants with corolla or petals finely dissected as in *Elaeocarpus* or deeply bifid as in *Dichapetalum*. The latter taxa are indicated by (2). In *Passifloraceae* it is the corona which is fimbriate.

Taxon	Family	Taxon	Family
Aceratium p.p.	Elaeoc.	*Ischnocarpus*	Apoc.
Anisophyllea	Rhiz.	*Kandelia*	Rhiz.
Bruguiera	Rhiz.	*Lophopetalum* p.p.	Celastr.
Carallia	Rhiz.	*Macaranga fimbriata* (Au)	Euph.
Caryophyllaceae (2)	Caryoph.	*Malpighia* p.p.*	Malp.
Ceriops	Rhiz.	*Nymphoides*	Gent.
Cocculus orbiculatus	Menisp.	*Olax* p.p.	Olacac.
Crispiloba (Au)	Alseu.	*Orchidaceae* p.p.	Orch.
Dichapetalum (2)	Dichap.	*Passifloraceae* (corona)	Passifl.
Dolichandrone spathacea	Bign.	*Rhizophora*	Rhiz.
Dubouzetia p.p.	Elaeoc.	*Scolopia* p.p.	Flac.
Elaeocarpus	Elaeoc.	*Sericolea* p.p. (2)	Elaeoc.
Erycibe (2)	Conv.	*Sloanea* p.p.	Elaeoc.
Euonymus p.p.	Celastr.	*Stereospermum fimbriatum*	Bign.
Gesneriaceae p.p.	Gesn.	*Thysanotus*	Liliac.
Gynotroches	Rhiz.	*Trichosanthes*	Cuc.
Hiptage p.p.	Malp.	*Trigonostemon diplopetalus* (2)	Euph.
Hodgsonia	Cuc.		

82. Corolla / petals with appendages — Fig. 52

Plants in which the corolla or petals bear appendages, as, e.g,. in many *Apocynaceae* and *Flacourtiaceae*.

Taxon	*Family*	*Taxon*	*Family*
Adenia	Passifl.	*Paropsia*	Flac.
Apocynaceae p.p.	Apoc.	*Passiflora*	Passifl.
Boraginaceae p.p.	Borag.	*Ryparosa*	Flac.
Cuscuta	Conv.	*Sabia*	Sab.
Erythroxylon	Erythr.	*Sapindaceae* p.p.	Sapind.
Hydnocarpus	Flac.	*Scaphocalyx*	Flac.
Meliosma	Sab.	*Thymelaeaceae* p.p.	Thym.
Pangium	Flac.	*Trichadenia*	Flac.

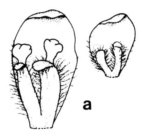

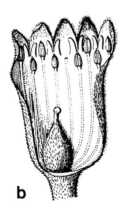

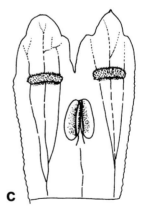

Figure 52. Corolla / petals with appendages – a. *Guioa pleuropteris* (Sapind.); b. *Enkleia malaccensis* (Thym.); c. *Cynoglossum javanicum* (Borag.).

83. Stamens opposite the petals

Plants in which the stamens are placed before the petals (e.g. *Rhamnaceae*) instead of alternating with them as is usually the case. Also plants where the stamens are opposite the tepals (petals absent).

Taxon	*Family*	*Taxon*	*Family*
Amaranthaceae	Amaran.	*Opiliaceae*	Opil.
Basellaceae	Basell.	*Papaveraceae*	Papav.
Berberidaceae	Berb.	*Phytolacca*	Phytol.
Chenopodiaceae	Chenop.	*Plumbaginaceae*	Plumb.
Corynocarpus	Coryn.	*Polygonaceae*	Polygon.
Crypteroniaceae	Crypter.	*Portulaccaceae*	Port.
Diospyros	Eben.	*Primulaceae*	Prim.
Dipterocarpaceae	Dipt.	*Proteaceae*	Prot.
Euphorbiaceae	Euph.	*Rhamnaceae*	Rhamn.
Flacourtiaceae	Flac.	*Rhizophoraceae*	Rhiz.
Loranthaceae	Loranth.	*Sabiaceae*	Sab.
Lythraceae	Lythr.	*Sapotaceae*	Sapot.
Melastomataceae	Melast.	*Sarcosperma*	Sarcosp.
Menispermaceae	Menisp.	*Sterculiaceae*	Sterc.
Myrsinaceae	Myrsin.	*Viscaceae*	Visc.
Olacaceae	Olacac.	*Vitaceae*	Vit.
Onagraceae	Onagr.		

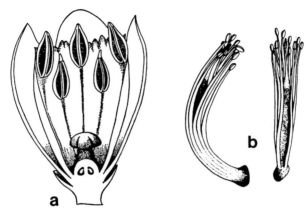

Figure 53. Staminal tube – a. *Reinwardtiodendron* (Meliac.); b. *Aeschynomene indica* (Legum.).

84. Staminal tube — Fig. 53

Stamens fused to form a tube, a very common feature of *Meliaceae*.

Taxon	Family	Taxon	Family
Bruinsmia	Styr.	*Nyctaginaceae*	Nyctag.
Camellia p.p.	Theac.	*Oxalidaceae*	Oxal.
Connaraceae p.p.	Connar.	*Polygalaceae* p.p.	Polygal.
Erythroxylon	Erythr.	*Rutaceae* p.p.	Rut.
Harmandia	Olacac.	*Sterculiaceae* p.p.	Sterc.
Leea	Leeac.	*Styrax*	Styr.
Leguminosae p.p.	Leg.	*Symplocos* p.p.	Sympl.
Linaceae	Linac.	*Tiliaceae* p.p.	Tiliac.
Malvaceae	Malv.	*Trigoniastrum*	Trigon.
Meliaceae p.p.	Meliac.	*Violaceae* p.p.	Viol.
Myrsinaceae p.p.	Myrsin.		

85. Stamens with appendages — Fig. 54

Plants in which the stamens bear hair tufts or scales on filaments or anthers, as e.g. in *Ericaceae* and *Icacinaceae*.

Taxon	Family	Taxon	Family
Agatea	Viol.	*Diplocyclos*	Cuc.
Asclepiadaceae	Asclep.	*Ecdysanthera*	Apoc.
Cantleya	Icac.	*Elaeocarpus*	Elaeoc.
Celastraceae	Celastr.	*Embolanthera*	Hamam.
Chloranthus	Chlor.	*Euphorbiaceae* p.p.	Euph.
Cinnamomum	Laur.	*Gaultheria*	Eric.
Compositae p.p.	Comp.	*Gomphandra*	Icac.
Dillenia	Dill.	*Harrisonia*	Simar.

(85. *Stamens with appendages, continued*)

Taxon	Family	Taxon	Family
Helicia	Prot.	*Polyalthia*	Annon.
Hybanthus p.p.	Viol.	*Premna*	Verb.
Indigofera	Leg.	*Rinorea*	Viol.
Justicia	Acanth.	*Rhyssopterys*	Malp.
Leviera	Monim.	*Stemona*	Stem.
Macrolenes	Melast.	*Stemonurus*	Icacin.
Madhuca	Sapot.	*Trichopus*	Diosc.
Magnolia	Magn.	*Typhonium*	Arac.
Medusanthera	Icac.	*Vaccinium*	Eric.
Meliosma	Sab.	*Viola*	Viol.
Munronia	Meliac.	*Zanthoxylum*	Rut.
Parashorea	Dipt.		

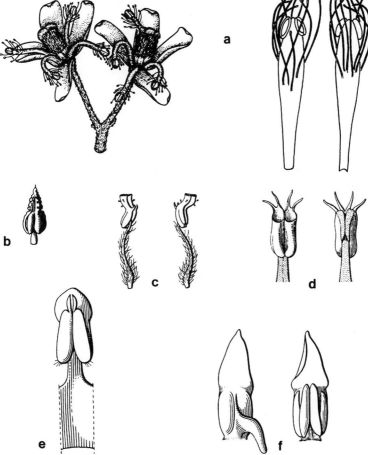

Figure 54. Stamens with appendages – a. *Gomphandra javanica*; b. *Kokoona ochracea* (Celastr.); c. *Vaccinium bancanum*; d. *Gaultheria punctata*; e. *Viola pilosa*; f. *Rinorea horneri*.

86. Anthers basifixed, apical pores — Fig. 55

The combination of basifixed anthers which open by apical pores is common in *Elaeocarpaceae* and *Ochnaceae*.

Taxon	Family
Aceratium	Elaeoc.
Argostemma	Rub.
Cassia	Leg.
Clematis	Ranunc.
Dillenia	Dill.
Elaeocarpus	Elaeoc.
Ericaceae p.p.	Eric.
Euthemis	Ochn.
Gomphia	Ochn.
Melastomataceae p.p.	Melast.
Myrsinaceae p.p.	Myrsin.
Ochna	Ochn.
Pentaphylax	Pentaph.
Solanaceae p.p.	Solan.
Tetracera	Dill.
Theaceae p.p.	Theac.
Wrightia	Apoc.

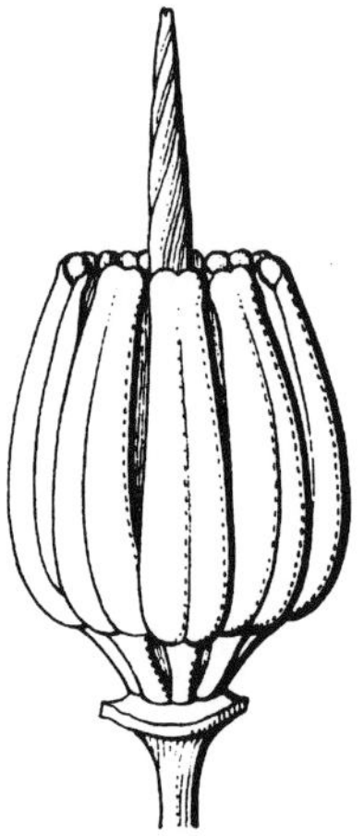

Figure 55. Anthers basifixed, apical pores – *Gomphia serrata*.

Figure 56. Anthers opening by valves – a. *Embolanthera spicata*; b. *Nothaphoebe umbelliflora*.

87. Anthers opening by valves — Fig. 56

Instead of opening by slits or pores the anthers open by one or more window-like structures; characteristic for *Lauraceae*.

Taxon	Family	Taxon	Family
Actinodaphne	Laur.	*Dryadodaphne*	Monim.
Actinolindera	Laur.	*Embolanthera*	Hamam.
Alseodaphne	Laur.	*Endiandra*	Laur.
Beilschmiedia	Laur.	*Eusideroxylon*	Laur.
Caryodaphnopsis	Laur.	*Gyrocarpus*	Hern.
Cinnadenia	Laur.	*Hernandia*	Hern.
Cinnamomum	Laur.	*Hexapora*	Laur.
Cryptocarya	Laur.	*Illigera*	Hern.
Dehaasia	Laur.	*Lindera*	Laur.

(87. Anthers opening by valves, continued)

Taxon	Family	Taxon	Family
Litsea	Laur.	*Polyporandra*	Icacin.
Neocinnamomum	Laur.	*Potoxylon*	Laur.
Neolitsea	Laur.	*Rhodoleia*	Hamam.
Nothaphoebe	Laur.	*Sycopsis*	Hamam.
Persea	Laur.	*Triadodaphne*	Laur.
Phoebe	Laur.		

88. Broad sessile stigma — Fig. 57

The ovary bears a broad flat stigma as seen in *Garcinia, Ilex* and others.

Taxon	Family	Taxon	Family
Aglaia p.p.	Meliac.	*Iodes*	Icacin.
Aphanamixis	Meliac.	*Kokoona*	Celastr.
Aporosa p.p.	Euph.	*Medusanthera*	Icacin.
Aquilaria	Thym.	*Miquelia*	Icacin.
Canarium p.p.	Burs.	*Octospermum*	Euph.
Cantleya	Icacin.	*Platea*	Icacin.
Champereia	Opil.	*Polyporandra*	Icacin.
Codiocarpus	Icacin.	*Pseudoclausena*	Meliac.
Dacryodes	Burs.	*Pyrenacantha*	Icacin.
Drypetes	Euph.	*Rhyticaryum*	Icacin.
Endospermum	Euph.	*Ryparosa*	Flac.
Erycibe	Conv.	*Santiria*	Burs.
Garcinia	Gutt.	*Sphenostemon*	Sphen.
Gomphandra	Icacin.	*Trimenia*	Trim.
Gonocaryum p.p.	Icacin.	*Triomma*	Burs.
Haplolobus	Burs.	*Walsura*	Meliac.
Hydnocarpus p.p.	Flac.	*Wikstroemia*	Thym.
Ilex	Aquif.		

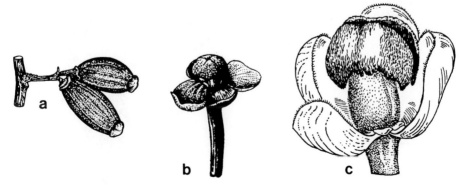

Figure 57. Broad sessile stigma – a. *Gomphandra javanica*; b. *Garcinia segmentata*; c. *Drypetes polyneura.*

89. Long forked style — Fig. 58

The style is divided to the base and the arms are divergent. Common in *Moraceae* and several *Euphorbiaceae*.

Taxon	Family	Taxon	Family
Aporosa p.p.	Euph.	*Nyssa*	Nyss.
Araliaceae p.p.	Aral.	*Polyosma*	Sax.
Buxus	Bux.	*Pteleocarpa*	Borag.
Cunoniaceae	Cun.	*Sapindaceae* p.p.	Sapind.
Daphniphyllum	Daphn.	*Sarcococca*	Bux.
Euphorbiaceae p.p.	Euph.	*Ulmaceae*	Ulm.
Hamamelidaceae p.p.	Hamam.	*Umbelliferae* p.p.	Umb.
Itea	Sax.	*Urticaceae* p.p.	Urt.
Moraceae p.p.	Morac.		

 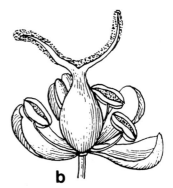

Figure 58. Long forked style – a. *Nyssa javanica*; b. *Celtis philippensis* (Ulm.).

90. Double forked style — Fig. 59

Like the previous but each arm of the style again divided, e.g. *Cordia*.

Taxon	Family
Aporosa p.p.	Euph.
Celtis	Ulm.
Cleidion	Euph.
Cleistanthus p.p.	Euph.
Cordia	Borag.
Croton p.p.	Euph.
Gelsemium	Logan.
Pteleocarpa	Borag.
Rhamnus	Rhamn.
Wetria	Euph.

Figure 59. Double forked style – *Aporosa lagenocarpa*.

91. Excentric style— Fig. 60

Plants in which the style is not terminal but basal or marginal. Common in *Sapindaceae* and *Sabiaceae*.

Taxon	Family	Taxon	Family
Antidesma p.p.	Euph.	*Menispermaceae* p.p.	Menisp.
Apodytes	Icacin.	*Ochnaceae*	Ochn.
Chrysobalanaceae	Chrysob.	*Pegia*	Anac.
Commelinaceae p.p.	Comm.	*Pimelea*	Thym.
Dichondra	Conv.	*Pleurostylia*	Celastr.
Dracontomelon	Anac.	*Ranunculaceae*	Ranunc.
Ficus p.p.	Morac.	*Sabia*	Sab.
Finschia	Prot.	*Santiria*	Burs.
Gluta	Anac.	*Sapindaceae* p.p.	Sapind.
Helicia	Prot.	*Spondias*	Anac.
Labiatae	Lab.	*Streblus*	Morac.
Mangifera	Anac.	*Suriana*	Simar.
Meliosma	Sab.		

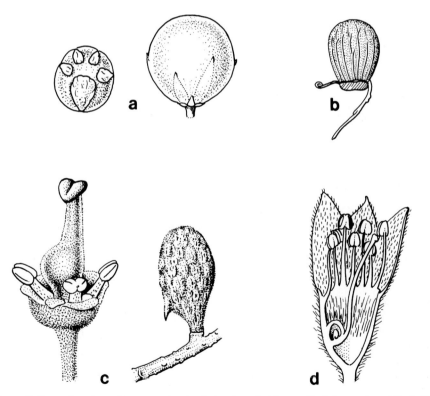

Figure 60. Excentric style – a. *Dracontomelon dao*; b. *Apodytes dimidiata*; c. *Nephelium maingayi* (Sapind.); d. *Parinari sumatrana* (Chrysob.).

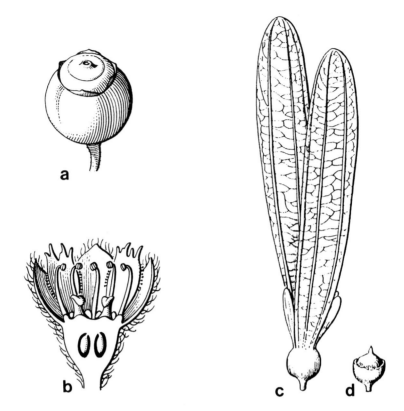

Figure 61. Ovary inferior – a. *Vaccinium bancanum*; b. *Anisophyllea disticha*; c. *Anisoptera grossivenia*; d. idem, wings removed.

92. Ovary inferior — Fig. 61

Ovary completely embedded in the hypanthium. Common in a few families such as *Rubiaceae* and *Caprifoliaceae*, exceptional in others such as *Dipterocarpaceae* and *Flacourtiaceae*. Taxa in which the ovary is incompletely inferior are indicated by (1).

Taxon	Family	Taxon	Family
Agapetes	Eric.	*Burmanniaceae*	Burm.
Alangium	Alang.	*Cactaceae* *	Cact.
Amaryllidaceae	Amaryll.	*Cannaceae* *	Cannac.
Anisophyllea	Rhiz.	*Caprifoliaceae*	Caprif.
Anisoptera (1)	Dipt.	*Carallia*	Rhiz.
Anneslea	Theac.	*Cassytha*	Laur.
Araliaceae	Aral.	*Ceriops* (1)	Rhiz.
Aristolochiaceae	Arist.	*Chloranthaceae*	Chlor.
Balanophoraceae	Balanoph.	*Codonopsis* (1)	Camp.
Begoniaceae	Begon.	*Combretocarpus*	Rhiz.
Bruguiera	Rhiz.	*Combretaceae*	Combr.

(92. Ovary inferior – continued)

Taxon	Family	Taxon	Family
Compositae	Comp.	*Myrtaceae* (not *Tristaniopsis*)	Myrt.
Corsia	Cors.	*Nymphaeaceae*	Nymph.
Costera	Eric.	*Nyssa*	Nyss.
Cryptocarya	Laur.	*Octomeles*	Datisc.
Cucurbitaceae	Cuc.	*Olacaceae* p.p.	Olacac.
Dimorphanthera	Eric.	*Onagraceae*	Onagr.
Dipterocarpus	Dipt.	*Orchidaceae*	Orch.
Engelhardia	Jugl.	*Pellacalyx*	Rhiz.
Eriobotrya	Rosac.	*Pentaphragma*	Pentapr.
Eupomatia	Eupom.	*Photinia*	Rosac.
Eusideroxylon	Laur.	*Potoxylon*	Laur.
Gardneria (1)	Logan.	*Punica* *	Punic.
Goodeniaceae	Good.	*Pyrus*	Rosac.
Gouania	Rhamn.	*Raphiolepis*	Rosac.
Haemodorum (1)	Haemod.	*Rhizophora*	Rhiz.
Haloragaceae	Halor.	*Rosa* *	Rosac.
Hamamelidaceae (1)	Hamam.	*Rubiaceae* (not *Gaertnera*)	Rub.
Hernandia	Hern.	*Ruthiella*	Camp.
Homalium (1)	Flac.	*Santalaceae*	Sant.
Kandelia	Rhiz.	*Saxifragaceae*, some (1)	Sax.
Laurentia *	Camp.	*Sciaphila*	Triur.
Lobelia	Camp.	*Sphenoclea*	Sphenoc.
Loranthaceae	Loranth.	*Stylidium*	Styl.
Lythraceae	Lythr.	*Tetragonia* (1)	Aizoac.
Maesa (1)	Myrsin.	*Tetrameles*	Datisc.
Malus *	Rosac.	*Triplostegia*	Dips.
Marantaceae	Marant.	*Umbelliferae*	Umb.
Mastixia	Corn.	*Vaccinium*	Eric.
Mastixiodendron (1)	Rub.	*Valeriana*	Val.
Melastomataceae	Melast.	*Viscaceae*	Visc.
Moraceae	Morac.	*Wahlenbergia* (1)	Camp.
Musaceae	Musac.	*Zingiberaceae*	Zing.

FRUIT (characters 93–101)

93. Fruits blue

Fruits ripening blue are exceptional. Common in *Elaeocarpus* and *Symplocos*.

Taxon	Family	Taxon	Family
Alyxia	Apoc.	*Lasianthus* p.p.	Rub.
Amaracarpus	Rub.	*Lepiniopsis*	Apoc.
Callicarpa p.p.	Verb.	*Litsea* p.p.	Laur.
Clidemia	Melast.	*Mastixia*	Corn.
Cryptocarya p.p.	Laur.	*Memecylon* p.p.	Melast.
Dianella	Liliac.	*Nertera* p.p.	Rub.
Dichroa	Sax.	*Peliosanthes*	Liliac.
Diplycosia	Eric.	*Phoebe* p.p.	Laur.
Disporum	Liliac.	*Pollia*	Comm.
Elaeocarpus p.p.	Elaeoc.	*Polygonum*	Polygon.
Erythropalum (seed)	Olacac.	*Polyosma*	Sax.
Euchresta	Leg.	*Psychotria* p.p.	Rub.
Eurya	Theac.	*Rubia*	Rub.
Harmandia	Olacac.	*Santiria* p.p.	Burs.
Helicia	Prot.	*Saprosma* p.p.	Rub.
Jasminum	Oleac.	*Symplocos* p.p.	Sympl.
Lantana *	Verb.	*Vaccinium* p.p.	Eric.

94. Woody fruits, scattered seeds — Fig. 62

Plants with large woody fruits, containing many scattered seeds as in most *Hydnocarpus* and *Xanthophyllum* species.

Taxon	Family	Taxon	Family
Aegle	Rut.	*Merrillia*	Rut.
Bertholletia *	Lecyth.	*Pimelodendron macrocarpum*	Euph.
Burkillanthus	Rut.	*Porterandia*	Rub.
Capparis	Capp.	*Rothmannia*	Rub.
Couroupita *	Lecyth.	*Salacia* p.p.	Celastr.
Crateva	Capp.	*Scaphocalyx*	Flac.
Crescentia *	Bign.	*Siphonodon*	Celastr.
Feronia (*Limonia*)	Rut.	*Strychnos*	Logan.
Gardenia	Rub.	*Urnularia*	Apoc.
Glennia	Sapind.	*Voacanga*	Apoc.
Hodgsonia	Cuc.	*Willughbeia*	Apoc.
Hydnocarpus	Flac.	*Xanthophyllum* p.p.	Polygal.
Melodinus	Apoc.	*Xylocarpus*	Meliac.

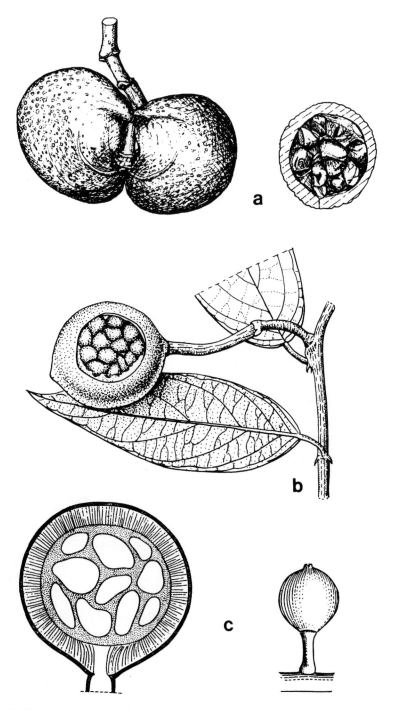

Figure 62. Woody fruits, scattered seeds – a. *Voacanga grandiflora*; b. *Capparis zeylanica*; c. *Hydnocarpus woodii*.

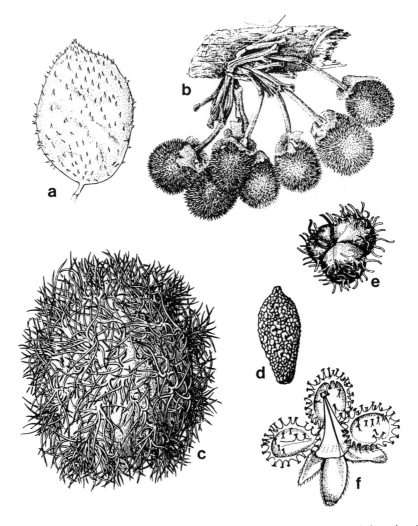

Figure 63. Spiny / muricate fruits – a. *Sindora velutina*; b. *Durio dulcis*; c. *Castanopsis hypophoenicia*; d. *Chilocarpus tuberculatus*; e. *Mallotus subpeltatus*; f. *Cynoglossum javanicum*.

95. Spiny / muricate fruits — Fig. 63

Fruits provided with soft processes such as *Nephelium*, stiff spiny ones such as *Castanopsis* or with a rugose surface such as *Parinari* or *Xerospermum*.

Taxon	Family	Taxon	Family
Acaena	Rosac.	*Ambrosia* p.p.	Comp.
Acanthospermum	Comp.	*Amomum* p.p.	Zing.
Agrimonia	Rosac.	*Annona* p.p. *	Annon.
Allamanda *	Apoc.	*Artocarpus*	Morac.
Amaranthaceae p.p.	Amaran.	*Asterostemma*	Asclep.

(95. Spiny / muricate fruits, continued)

Taxon	Family	Taxon	Family
Bidens	Comp.	*Martynia* *	Pedal.
Bixa *	Bixac.	*Melanochyla* p.p.	Anac.
Boraginaceae p.p.	Borag.	*Melastoma beccarianum*	Melast.
Byttneria	Sterc.	*Microdesmis*	Euph.
Caesalpinia p.p.	Leg.	*Mimosa* *	Leg.
Caldesia p.p.	Alism.	*Momordica* p.p.	Cuc.
Castanopsis	Fagac.	*Monocarpia* p.p.	Annon.
Cephalomappa	Euph.	*Muellerargia*	Cuc.
Ceratophyllum	Cerat.	*Myrica*	Myric.
Ceuthostoma	Casuar.	*Neesia*	Bomb.
Chaetocarpus	Euph.	*Nephelium*	Sapind.
Chilocarpus tuberculatus	Apoc.	*Omphalodes* p.p.	Borag.
Chionanthus pluriflorus	Oleac.	*Opuntia* *	Cact.
Chlaenandra	Menisp.	*Ormocarpum*	Leg.
Clappertonia *	Tiliac.	*Pandanus*	Pand.
Coelostegia	Bomb.	*Parabaena*	Menisp.
Commersonia	Sterc.	*Paranephelium*	Sapind.
Corchorus	Tiliac.	*Parartocarpus*	Morac.
Cosmos p.p.	Comp.	*Pimelodendron macrocarpum*	Euph.
Cubilia	Sapind.	*Praravinia verruculosa*	Rub.
Cullenia	Bomb.	*Priva* *	Verb.
Cyanandrium	Melast.	*Pseuduvaria*	Annon.
Cynanchum	Asclep.	*Pternandra*	Melast.
Cynoglossum	Borag.	*Ptychopyxis caput-medusae*	Euph.
Cyclanthera	Cuc.	*Ranunculus*	Ranunc.
Delphyodon	Apoc.	*Ricinus* p.p.	Euph.
Dichapetalum p.p.	Dichap.	*Rinorea anguifera*	Viol.
Dimocarpus p.p.	Sapind.	*Sagittaria* p.p.	Alism.
Dimorphocalyx muricatus	Euph.	*Salomonia*	Polygal.
Durio	Bomb.	*Sanicula*	Umb.
Ecballium	Cuc.	*Schleichera*	Sapind.
Erythrospermum	Flac.	*Schrankia*	Leg.
Euonymus p.p.	Celastr.	*Sebastiania* p.p.	Euph.
Fittingia p.p.	Myrsin.	*Sida* p.p.	Malv.
Flindersia	Rut.	*Sindora* p.p.	Leg.
Freycinetia	Pand.	*Sloanea* p.p.	Elaeoc.
Glossogyne	Comp.	*Spathiostemon*	Euph.
Gomphocarpus *	Asclep.	*Taxillus*	Loranth.
Gramineae p.p.	Gram.	*Trapa*	Trap.
Hydnocarpus polypetala	Flac.	*Tribulus*	Zygoph.
Jarandersonia	Tiliac.	*Trichosanthes* p.p.	Cuc.
Josephinia	Pedal.	*Triumfetta*	Tiliac.
Kostermansia	Bomb.	*Umbelliferae* p.p.	Umb.
Lasiococca	Euph.	*Uncinia*	Cyp.
Litchi	Sapind.	*Urena*	Malv.
Lithocarpus	Fagac.	*Xanthium* p.p.	Comp.
Macaranga p.p.	Euph.	*Xerospermum* p.p.	Sapind.
Macrolenes	Melast.	*Zippelia*	Piper.
Mallotus p.p.	Euph.	*Zornia*	Leg.

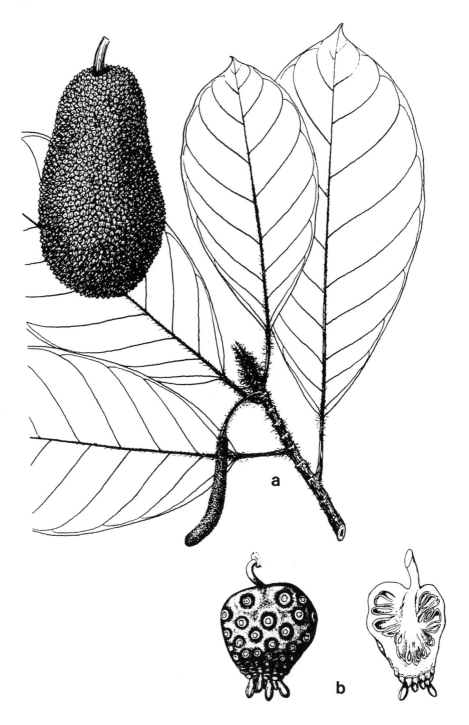

Figure 64. Compound fruits – a. *Artocarpus integer*; b. *Morinda citrifolia*.

96. Compound fruits — Fig. 64

Several fruits connate into a single structure as exemplified by *Artocarpus, Rubus* and *Nauclea.*

Taxon	Family	Taxon	Family
Adina	Rub.	*Maclura*	Morac.
Agathis	Arauc.	*Malaisia*	Morac.
Altingia	Hamam.	*Maoutia*	Urt.
Anakasia	Aral.	*Meryta* (P)	Aral.
Annanas *	Brom.	*Metadina*	Rub.
Annona *	Annon.	*Morinda*	Rub.
Anthocephalus	Rub.	*Morus*	Morac.
Antiaris	Morac.	*Nauclea*	Rub.
Antiaropsis	Morac.	*Osmoxylon*	Aral.
Araceae	Arac.	*Pandanus*	Pand.
Araucaria	Arauc.	*Parartocarpus*	Morac.
Artocarpus	Morac.	*Peperomia*	Piper.
Astrothalamus	Urt.	*Phytocrene*	Icacin.
Banksia	Prot.	*Pinus*	Conif.
Broussonetia	Morac.	*Piper*	Piper.
Casuarina	Casuar.	*Poikilospermum*	Urt.
Ceuthostoma	Casuar.	*Potentilla*	Rosac.
Coelospermum	Rub.	*Pothomorphe* *	Piper.
Cunoniaceae p.p.	Cun.	*Prainea*	Morac.
Dendrocnide	Urt.	*Procris*	Urt.
Elatostema	Urt.	*Rennellia*	Rub.
Etlingera	Zing.	*Rhodoleia*	Hamam.
Ficus	Morac.	*Rollinia* *	Annon.
Freycinetia	Pand.	*Rubus*	Rosac.
Gymnostoma	Casuar.	*Sararanga*	Pand.
Hullettia	Morac.	*Schefflera* p.p.	Aral.
Kadsura	Schis.	*Schisandra*	Schis.
Kibara	Monim.	*Streblus* p.p.	Morac.
Leucosyke	Urt.	*Zingiberaceae* p.p.	Zing.

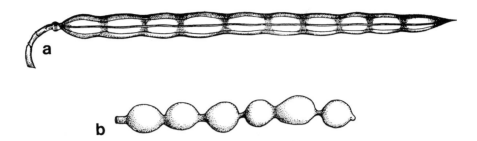

Figure 65. Moniliform fruit – a. *Moringa oleifera*; b. *Chilocarpus conspicuus.*

97. Moniliform fruit — Fig. 65

A usually elongated fruit constricted at intervals and giving the fruit the appearance of
a string of beads, e.g. *Sophora, Alyxia.*

Taxon	Family	Taxon	Family
Acacia p.p.	Leg.	*Neokeithia*	Apoc.
Airyantha	Leg.	*Ormocarpum*	Leg.
Alyxia p.p.	Apoc.	*Orophea*	Annon.
Chilocarpus	Apoc.	*Parameria*	Apoc.
Cleghornia	Apoc.	*Parkinsonia* *	Leg.
Dasymaschalon	Annon.	*Pottsia*	Apoc.
Derris p.p.	Leg.	*Rauwenhoffia*	Annon.
Desmodium p.p.	Leg.	*Rhodomyrtus* p.p.	Myrt.
Desmos	Annon.	*Sophora*	Leg.
Erythrina p.p.	Leg.	*Tamarindus*	Leg.
Friesodielsia p.p.	Annon.	*Urceola*	Apoc.
Hollarrhena	Apoc.	*Xylopia* p.p.	Annon.
Moringa *	Moring.		

98. Fruit winged — Fig. 66 (see also Fig. 32, p. 74)

Fruits provided by flat structures of different origin: in *Dipterocarpaceae* the wings
are formed by accrescent calyx lobes, in *Engelhardia* the wings are formed by bracts,
in *Combretum* the fruit is provided with thin ridges and in *Pterocarpus* the fruit is flat.

Taxon	Family	Taxon	Family
Acer	Acer.	*Baccaurea angulata*	Euph.
Ailanthus	Simar.	*Bauhinia scandens*	Leg.
Ancistrocladus	Ancistr.	*Begonia*	Begon.
Anisoptera	Dipt.	*Berrya*	Tiliac.
Argyrodendron (Au)	Sterc.	*Brachylophon*	Malp.
Aspidopteris	Malp.	*Butea*	Leg.
Atalaya	Sapind.	*Callitriche*	Callitr.

(98. Fruit winged, continued)

Taxon	Family	Taxon	Family
Calycopteris	Combr.	*Derris*	Leg.
Cardiopteris	Card.	*Dioscorea*	Diosc.
Ceratopetalum virchowii (Au)	Cun.	*Dipterocarpus*	Dipt.
Colona	Tiliac.	*Dodonaea*	Sapind.
Combretocarpus	Rhiz.	*Dryobalanops*	Dipt.
Combretodendron	Lecyth.	*Engelhardia*	Jugl.
Combretum	Combr.	*Erythrina* p.p.	Leg.
Congea	Verb.	*Firmiana*	Sterc.
Cotylelobium	Dipt.	*Fraxinus*	Oleac.
Dalbergia	Leg.	*Gillbeea*	Cun.

→

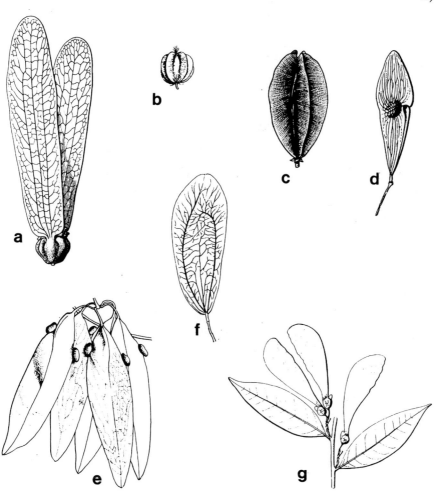

Figure 66. Fruit winged – a. *Dipterocarpus cornutus*; b. *Pentace excelsa*; c. *Lophopyxis maingayi*; d. *Ailanthus excelsa*; e. *Firmiana malayana*; f. *Koompassia malaccensis*; g. *Securidaca ecristata*.

(98. Fruit winged, continued)

Taxon	Family	Taxon	Family
Glochidion p.p.	Euph.	*Plagiopteron* (As)	Plag.
Gluta p.p.	Anac.	*Porana*	Conv.
Gouania	Rhamn.	*Pteleocarpa*	Borag.
Guioa	Sapind.	*Pterocarpus*	Leg.
Gyrocarpus	Hern.	*Pterococcus*	Euph.
Harmandia	Olacac.	*Pterocymbium*	Sterc.
Hedyotis pterita	Rub.	*Pterolobium*	Leg.
Heritiera p.p.	Sterc.	*Quisqualis*	Combr.
Hernandia p.p.	Hern.	*Rhyssopterys*	Malp.
Hildegardia	Sterc.	*Samadera*	Simar.
Hiptage	Malp.	*Sarcopteryx*	Sapind.
Hopea	Dipt.	*Scaphium*	Sterc.
Hugonia	Linac.	*Schoutenia*	iliac.
Hymenocardia	Euph.	*Securidaca*	Polygal.
Illigera	Hern.	*Shorea*	Dipt.
Jackiopsis	Rub.	*Smythea*	Rhamn.
Kalappia	Leg.	*Soulamea*	Simar.
Kleinhovia	Sterc.	*Spatholobus*	Leg.
Koompassia	Leg.	*Sphenodesme*	Verb.
Kydia (As)	Malv.	*Steenisia*	Rub.
Lophopyxis	Loph.	*Stenomeris*	Diosc.
Macaranga p.p.	Euph.	*Sterculia laurifolia* (Au)	Sterc.
Macropteranthes (Au)	Combr.	*Storckiella* (Au P)	Leg.
Mallotus sumatranus	Euph.	*Swintonia*	Anac.
Marsdenia p.p.	Asclep.	*Symphorema*	Verb.
Maxwellia (P)	Sterc.	*Terminalia* p.p.	Combr.
Megistostigma burmannicum	Euph.	*Tetractomia*	Rut.
Myriopteron	Asclep.	*Trigoniastrum*	Trigon.
Neobalanocarpus	Dipt.	*Triomma*	Burs.
Neuropeltis	Conv.	*Tripterygium* (As)	Celastr.
Neuropeltopsis	Conv.	*Tristellateia*	Malp.
Pajanelia	Bign.	*Tristira*	Sapind.
Parashorea	Dipt.	*Ulmus*	Ulm.
Parishia	Anac.	*Ungeria* (P)	Sterc.
Pentace	Tiliac.	*Upuna*	Dipt.
Peripterygia (P)	Celastr.	*Vatica* p.p.	Dipt.
Petraeovitex	Verb.	*Ventilago*	Rhamn.
Petrea *	Verb.	*Zollingeria*	Sapind.

99. Fruit ridged — Fig. 67

Fruits provided with (usually longitudinal) ridges; when very conspicuously raised they are considered winged fruits. Example of ridged fruits: *Helicia, Myristicaceae.*

Taxon	Family	Taxon	Family
Alangium p.p.	Alang.	*Gonocaryum* p.p.	Icacin.
Allantospermum	Simar.	*Helicia* p.p.	Prot.
Annonaceae p.p.	Annon.	*Hernandia* p.p.	Hern.
Apodytes	Icacin.	*Leguminosae* p.p.	Leg.
Baccaurea trigonocarpa	Euph.	*Macadamia*	Prot.
Barringtonia p.p.	Lecyth.	*Mallotus* p.p.	Euph.
Boerhavia	Nyctag.	*Manihot esculenta* *	Euph.
Burseraceae p.p.	Burs.	*Meliosma*	Sab.
Campanulaceae p.p.	Camp.	*Myristicaceae* p.p.	Myrist.
Casearia p.p.	Flac.	*Pentastemona*	Pent.
Chionanthus p.p.	Oleac.	*Phytocrene* p.p.	Icacin.
Connaraceae p.p.	Connar.	*Psychotria*	Rub.
Cryptocarya p.p.	Laur.	*Ptychopyxis costata*	Euph.
Dichapetalum p.p.	Dichap.	*Quassia* p.p.	Simar.
Dregea	Asclep.	*Scyphostegia*	Scyph.
Dysoxylum caulostachyum	Meliac.	*Sterculiaceae* p.p.	Sterc.
Euphorbiaceae p.p.	Euph.	*Terminalia* p.p.	Combr.
Finlaysonia	Asclep.	*Teijsmanniodendron* p.p.	Verb.
Garcinia p.p.	Gutt.	*Thevetia* *	Apoc.
Gomphandra p.p.	Icacin.	*Timonius* p.p.	Rub.

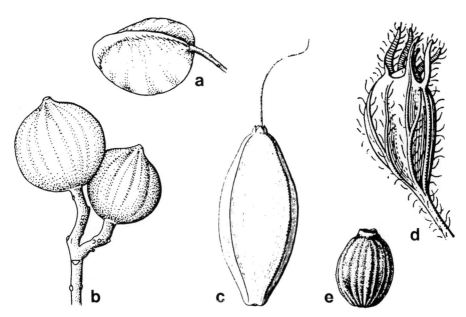

Figure 67. Fruit ridged – a. *Heritiera littoralis* (Sterc.); b. *Cryptocarya densiflora*; c. *Barringtonia macrostachys*; d. *Ruthiella saxicola* (Camp.); e. *Alangium ridleyi.*

100. Lagerstroemia capsule — Fig. 68

A more or less round capsule splitting at the top as in *Lagerstroemia, Metrosideros* and *Schima*.

Taxon	*Family*
Axinandra	Crypter.
Coelostegia	Bomb.
Cratoxylum	Gutt.
Crypteronia	Crypter.
Dactylocladus	Crypter.
Distylium	Hamam.
Duabanga	Sonn.
Dubouzetia	Elaeoc.
Gordonia	Theac.
Ixonanthes	Linac.
Lagerstroemia	Lythr.
Leptospermum	Myrt.
Maingaya	Hamam.
Metrosideros	Myrt.
Neesia	Bomb.
Rhodoleia	Hamam.
Schima	Theac.
Sloanea	Elaeoc.
Sycopsis	Hamam.
Tristaniopsis	Myrt.
Xanthostemon	Myrt.

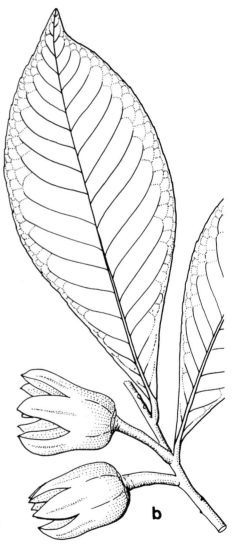

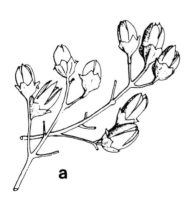

Figure 68. Lagerstroemia capsule – a. *Lagerstroemia floribunda*; b. *Gordonia grandiflora*.

101. Three-locular capsule — Fig. 69 (see also Fig. 63e, p. 124)

Most *Euphorbiaceae* have this type of fruit, but it is known in several other families, such as *Theaceae* and *Celastraceae*.

Taxon	Family	Taxon	Family
Acalypha	Euph.	*Chondrostylis*	Euph.
Actephila	Euph.	*Cladogynos*	Euph.
Agatea	Viol.	*Claoxylon* p.p.	Euph.
Agrostistachys	Euph.	*Cleidion* p.p.	Euph.
Alchornea	Euph.	*Cleistanthus*	Euph.
Allantospermum	Simar.	*Cnesmone* p.p.	Euph.
Amaryllidaceae	Amaryll.	*Colubrina*	Rhamn.
Amesiodendron p.p.	Sapind.	*Croton* p.p.	Euph.
Aporosa p.p.	Euph.	*Dichapetalum*	Dichap.
Arthropodium	Liliac.	*Dicoelia*	Euph.
Asthonia	Euph.	*Dimorphocalyx*	Euph.
Austrobuxus	Euph.	*Elateriospermum*	Euph.
Baccaurea p.p.	Euph.	*Emmenosperma*	Rhamn.
Blachia	Euph.	*Epiprinus*	Euph.
Blumeodendron p.p.	Euph.	*Erismanthus*	Euph.
Boesenbergia	Zing.	*Erythrospermum*	Flac.
Botryophora	Euph.	*Euphorbia*	Euph.
Brachychilum	Zing.	*Excoecaria*	Euph.
Breynia	Euph.	*Fahrenheitia*	Euph.
Caesia	Liliac.	*Flueggea*	Euph.
Camellia	Theac.	*Glochidion* p.p.	Euph.
Canna *	Cannac.	*Gloriosa*	Liliac.
Casearia	Flac.	*Gonystylus* p.p.	Thym.
Celastrus	Celastr.	*Guioa*	Sapind.
Cephalomappa	Euph.	*Harpullia* p.p.	Sapind.
Chaetocarpus	Euph.	*Hedychium*	Zing.
Cheilosa	Euph.	*Hevea* *	Euph.
Chlorophytum	Liliac.	*Homonoia*	Euph.

→

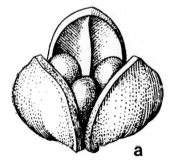 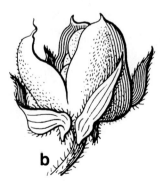

Figure 69. Trilocular capsule – a. *Gonystylus bancanus*; b. *Viola pilosa*.

(101. Three-locular capsule, continued)

Taxon	Family	Taxon	Family
Hybanthus	Viol.	*Reissantia*	Celastr.
Iphigenia	Liliac.	*Richeriella*	Euph.
Koilodepas	Euph.	*Rinorea*	Viol.
Kokoona	Celastr.	*Sapium* p.p.	Euph.
Lepisanthes p.p.	Sapind.	*Sarcococca*	Bux.
Leptopus	Euph.	*Sauropus*	Euph.
Lilium	Liliac.	*Sebastiania*	Euph.
Lophopetalum	Celastr.	*Spathiostemon*	Euph.
Macaranga p.p.	Euph.	*Sumbaviopsis*	Euph.
Mallotus	Euph.	*Suregada*	Euph.
Margaritaria	Euph.	*Synostemon*	Euph.
Maytenus	Celastr.	*Thysanotus*	Liliac.
Melanolepis	Euph.	*Tricyrtis*	Liliac.
Osmelia	Flac.	*Trigonachras*	Sapind.
Paranephelium p.p.	Sapind.	*Trigonopleura*	Euph.
Petrosavia	Liliac.	*Trigonostemon*	Euph.
Phyllanthus p.p.	Euph.	*Viola*	Viol.
Pittosporum	Pitt.	*Wetria*	Euph.
Ptychopyxis	Euph.	*Zingiber* p.p.	Zing.

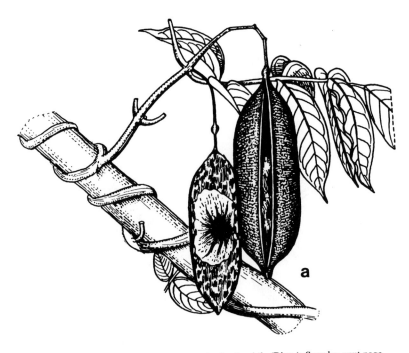

Figure 70. Seeds winged – a. *Tecomanthe dendrophila* (Bign.). See also next page.

SEED (characters 102–105)

102. Seeds winged — Fig. 70 (see also Fig. 31, p. 72)

Seeds with a thin flat appendage as in *Casuarinaceae* and *Bignoniaceae*.

Taxon	Family	Taxon	Family
Acsmithia	Cun.	*Schima*	Theac.
Aganosma	Apoc.	*Schrebera*	Oleac.
Agatea	Viol.	*Schuurmansia*	Ochn.
Agathis	Arauc.	*Spiraeanthemum*	Cun.
Alloxylon	Prot.	*Stenomeris*	Diosc.
Altingia	Hamam.	*Swietenia* *	Meliac.
Amaryllidaceae p.p.	Amaryll.	*Tetractomia*	Rut.
Aristolochia	Arist.	*Toona*	Meliac.
Banksia p.p.	Prot.	*Triomma*	Burs.
Bignoniaceae p.p.	Bign.	*Tristaniopsis*	Myrt.
Bikkia	Rub.	*Tylophora*	Asclep.
Caldcluvia p.p.	Cun.	*Uncaria*	Rub.
Casuarina	Casuar.	*Wendlandia*	Rub.
Ceuthostoma	Casuar.	*Wightia*	Scroph.
Cinchona *	Rub.		
Coptosapelta	Rub.		
Cratoxylum	Gutt.		
Crypteronia	Crypter.		
Dactylocladus	Crypter.		
Dioscorea	Diosc.		
Eucryphia (Au)	Euph.		
Flindersia	Rut.		
Gelsemium	Logan.		
Gordonia	Theac.		
Grevillea	Prot.		
Gymnostoma	Casuar.		
Hymenodictyon	Rub.		
Hymenosporum	Pitt.		
Itoa	Flac.		
Ixonanthes	Linac.		
Kokoona	Celastr.		
Lagerstroemia	Lythr.		
Liliaceae p.p.	Liliac.		
Loeseneriella	Celastr.		
Lophopetalum	Celastr.		
Macrozanonia	Cuc.		
Moringa *	Moring.		
Mussaendopsis	Rub.		
Neonauclea	Rub.		
Pinus	Pinac.		
Pterospermum	Sterc.		
Pterygota	Sterc.		
Reissantia	Celastr.		
Rinorea p.p.	Viol.		

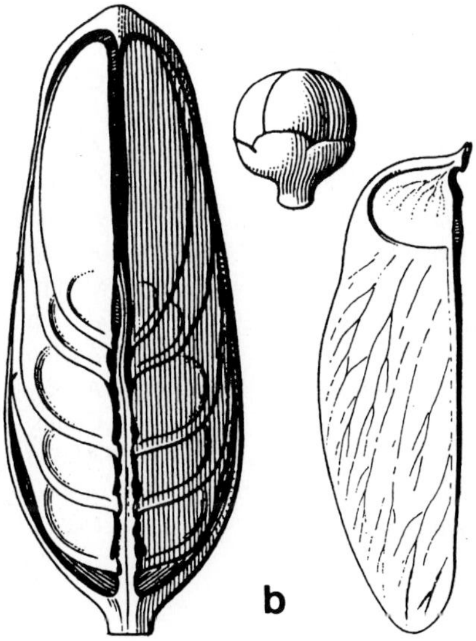

Figure 70. Seeds winged – a. *Tecomanthe dendrophila* (see previous page); b. *Kokoona ovatolanceolata*.

103. Seeds comose— Fig. 71

Seeds provided with a tuft of hairs such as in many *Apocynaceae, Asclepiadaceae* and *Compositae*.

Taxon	Family	Taxon	Family
Aeschynanthus	Gesn.	*Laggera*	Comp.
Aganosma	Apoc.	*Launaea*	Comp.
Alstonia	Apoc.	*Marsdenia*	Asclep.
Anaphalis	Comp.	*Micrechites*	Apoc.
Anodendron	Apoc.	*Microglossa*	Comp.
Asclepias *	Asclep.	*Microstemma*	Asclep.
Atherandra	Asclep.	*Mikania*	Comp.
Blumea	Comp.	*Nerium* *	Apoc.
Calotropis	Asclep.	*Parameria*	Apoc.
Ceropegia	Asclep.	*Parsonsia*	Apoc.
Chonemorpha	Apoc.	*Phyllanthera*	Asclep.
Cochlospermum	Cochl.	*Physostelma*	Asclep.
Conyza	Comp.	*Pluchea*	Comp.
Crassocephalum	Comp.	*Pottsia*	Apoc.
Cryptolepis	Asclep.	*Pterocaulon*	Comp.
Cryptostegia	Asclep.	*Raphistemma*	Asclep.
Cynanchum	Asclep.	*Rhynchospermum*	Comp.
Dischidia	Asclep.	*Salix*	Salic.
Dregea	Asclep.	*Sarawakodendron*	Celast.
Ecdysanthera	Apoc.	*Sarcostemma*	Asclep.
Emilia	Comp.	*Secamone*	Asclep.
Epilobium	Onagr.	*Senecio*	Comp.
Erechtites	Comp.	*Sonchus*	Comp.
Erigeron	Comp.	*Stephanotis*	Asclep.
Eupatorium	Comp.	*Streptocaulon*	Asclep.
Finlaysonia	Asclep.	*Strophanthus*	Apoc.
Genianthus	Asclep.	*Telosma*	Asclep.
Gnaphalium	Comp.	*Tetramolopium*	Comp.
Gymnanthera	Asclep.	*Toxocarpus*	Asclep.
Gymnema	Asclep.	*Trachelospermum*	Apoc.
Gynura	Comp.	*Tylophora*	Asclep.
Heterostemma	Asclep.	*Urceola*	Apoc.
Hoya	Asclep.	*Vallaris*	Apoc.
Ichnocarpus	Apoc.	*Vernonia*	Comp.
Inula	Comp.	*Weinmannia*	Cun.
Ischnostemma	Asclep.	*Wrightia*	Apoc.
Kibatalia	Apoc.	*Youngia*	Comp.
Lactuca	Comp.		

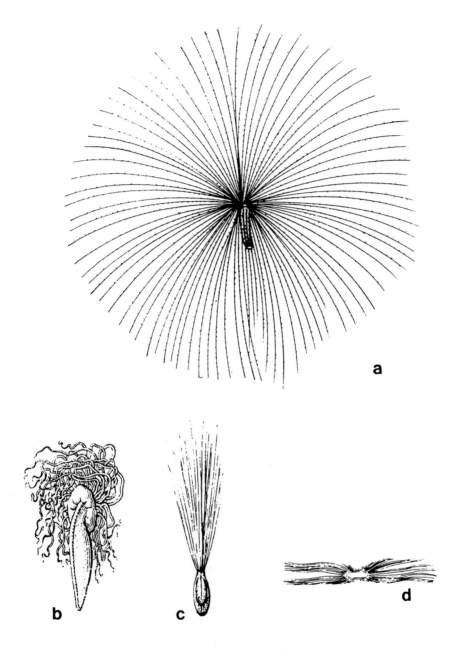

Figure 71. Seeds comose – a. *Crassocephalum crepidioides*; b. *Sarawakodendron filamentosum*; c. *Asclepias curassavica*; d. *Alstonia spathulata*.

104. Seeds arillate — Fig. 72

A usually fleshy and coloured outgrowth of the funicle surrounding the seed, as e.g. in *Meliaceae, Myristicaceae* and *Sapindaceae.*

Taxon	Family	Taxon	Family
Annonaceae p.p.	Annon.	*Thymelaeaceae* p.p.	Thym.
Apocynaceae p.p.	Apoc.	*Violaceae*	Viol.
Bombacaceae p.p.	Bomb.	*Zingiberaceae*	Zing.
Celastraceae	Celastr.		
Commelinaceae	Comm.		
Coniferae p.p.	Conif.		
Connaraceae	Connar.		
Dilleniaceae	Dill.		
Dubouzetia	Elaeoc.		
Euphorbiaceae p.p.	Euph.		
Flacourtiaceae p.p.	Flac.		
Guttiferae p.p.	Gutt.		
Leguminosae p.p.	Leg.		
Linaceae p.p.	Linac.		
Magnoliaceae	Magn.		
Marantaceae	Marant.		
Meliaceae p.p.	Meliac.		
Musaceae	Musac.		
Myristicaceae	Myrist.		
Oxalidaceae	Oxal.		
Papaveraceae	Papav.		
Passifloraceae	Passifl.		
Polygalaceae p.p.	Polygal.		
Sapindaceae	Sapind.		
Sloanea	Elaeoc.		

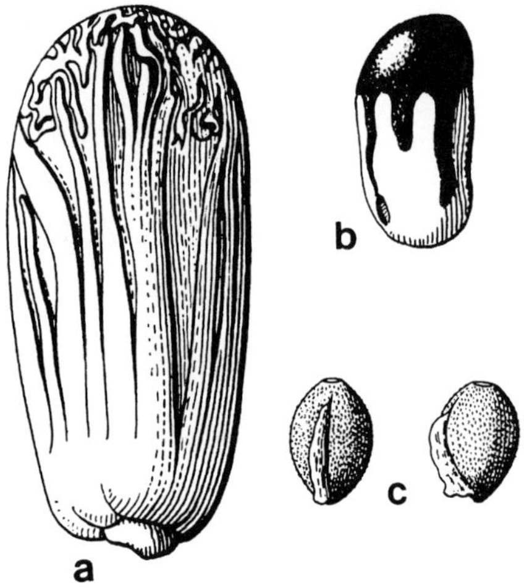

Figure 72. Seeds arillate – a. *Myristica papyracea*; b. *Ellipanthus tomentosus* (Connar.); c. *Viola pilosa*.

105. Ruminate endosperm — Fig. 73

The endosperm of the seeds is folded and on cross section looks like brains. Common in *Annonaceae* and *Myristicaceae.*

Taxon	Family	Taxon	Family
Alyxia	Apoc.	*Erycibe*	Conv.
Annonaceae	Annon.	*Fagaceae* p.p.	Fagac.
Araliaceae p.p.	Aral.	*Gonocaryum* p.p.	Icacin.
Arcangelisia	Menisp.	*Kostermanthus*	Chrys.
Atuna	Chrys.	*Leea*	Leeac.
Diospyros p.p.	Eben.	*Lepiniopsis*	Apoc.
Discocalyx p.p.	Myrsin.	*Loheria*	Myrsin.
Elaeocarpus p.p.	Elaeoc.	*Mangifera*	Anacard.

(105. Ruminate endosperm, continued)

Taxon	Family	Taxon	Family
Myristicaceae p.p.	Myrist.	*Tiliacora*	Menisp.
Palmae p.p.	Palm.	*Tinospora*	Menisp.
Polyosma p.p.	Sax.	*Trichopus*	Diosc.
Tabernaemontana	Apoc.	*Trimenia*	Trim.
Tapeinosperma p.p.	Myrsin.	*Viburnum* p.p.	Caprif.
Tetramerista	Theac.	*Voacanga*	Apoc.
Tetrastigma	Vit.		

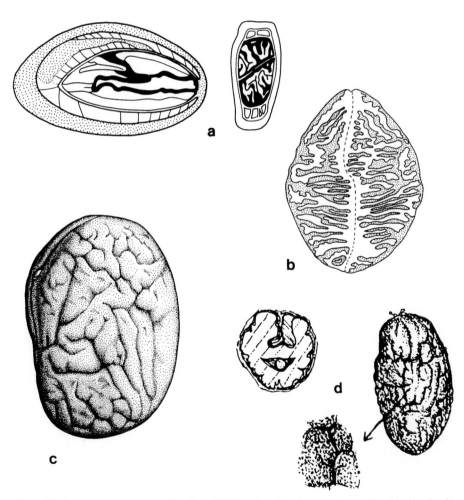

Figure 73. Ruminate endosperm – a. *Erycibe griffithii*; b. *Mangifera inocarpoides*; c. *Mangifera havilandii*; d. *Voacanga grandiflora*.

INDEX OF TAXA

All names mentioned in the 105 lists have been indexed and referred to page numbers; names mentioned in the figure legends have been marked with an asterisk.